Neutron Stars and NuSTAR

A Systematic Survey of Neutron Star Masses in
High Mass X-ray Binaries
and
Characterization of CdZnTe Detectors for NuSTAR

Varun B. Bhalerao

DISSERTATION.COM

Boca Raton

Neutron Stars and NuSTAR:
A Systematic Survey of Neutron Star Masses in High Mass X-ray Binaries and
Characterization of CdZnTe Detectors for NuSTAR

Dissertation.com
Boca Raton, Florida
USA • 2012

ISBN-10: 1-61233-944-1
ISBN-13: 978-1-61233-944-3

Cover image credits: (Top) Artist's rendering of Cygnus X-1, a black
hole high mass X-ray binary, courtesy of ESA/Hubble; (Bottom) Artist's
rendering of the NuSTAR satellite in space, courtesy of NuSTAR team.

Dedicated to my parents,
Erawati and Bhalchandra Bhalerao
my role models, motivation,
and strength.

Acknowledgments

My six years at Caltech have been an amazing experience, in both work and personal life. As I write this thesis summarizing the work, it is fitting that I begin by thanking the many people who have made my little achievement possible.

I am very fortunate to have Professors Fiona Harrison and Shri Kulkarni as my advisors. Their foresight, work ethic, and complementary styles of mentoring provided all the ingredients necessary to take me through my thesis. They have provided all resources I needed, and key insights whenever I was short on ideas (or had too many). I have learned a lot from working with them, and will continue striving to live up to their standards.

Working on NuSTAR has been a dream come true. No, really. Who has not wanted to build and launch a satellite? My work would have been impossible without significant help from my lab partners Brian Grefenstette and Vikram Rana. As the junior-most member in the Caltech NuSTAR team, I was the least experienced. I am grateful to Eric Bellm, Jill Burnham, Rick Cook, Takao Kitaguchi, Kristin Madsen, Peter Mao, and Hiromasa Miyasaka for helping me whenever I was stuck. I thank Peter's dentist. Miles Robinson, Steve Stryker, and Robert Crabill have been wonderful troubleshooters and the go-to people for any work in the laboratory. Summer projects by students Nancy Wu and Suk Sien Tie laid good foundations for my future work.

As an electrical engineer, I had a lot to learn about astronomy. My foray into observations was possible thanks to the enthusiastic guidance of Mansi Kasliwal. I thank Professor Marten van Kerkwijk for teaching me how to extract every last bit of information from data—a skill I still strive to perfect. Brad Cenko was extremely helpful in my first year photometer project. I thank Mike Muno for his early work on radial velocities of high mass X-ray binaries, which developed into my observational project. My batchmates Laura, Walter, Matthew, and Yacine have been a constant source of support and inspiration at Caltech. My work would have been nowhere as much fun without my long-time officemates Gwen and Laura. Together with Mansi, Shriharsh, Swarnima, and Abhilash, they made work a lot of fun! Shriharsh was the sounding board for new ideas at work. I am glad to have the company of Swarnima, Ryan and Krzysztof to share my passion of outreach and amateur astronomy.

Part of what makes Caltech special is the excellent administrative and technical staff, whose success lies in the fact that they are always in the background. My first-year project and subsequent observing runs and Palomar were a pleasure thanks to Kevin Rykoski, Dipali, and Jean Mueller. I thank Richard Dekany, Dan McKenna, and Jeff Zolkower for support on my first year project. The ADPF group, especially Patrick Shopbell and Anu Mahabal, have maintained an excellent computing infrastructure—giving me freedom from install logs and core dumps. Librarians Lindsay Cleary and Joy Painter always went out of their way to help with procuring and issuing books. Special thanks to the several administrative assistants at Caltech: Debby Miles, Nina Borg, and Caprece Anderson at the Space Radiation Laboratory, Bronagh Glaser at Geology and Planetary Sciences, Gina Armas and Gita Patel in Robinson and Cahill, and Teresita Legaspi at the Registrar's Office. Their untiring efforts ensured that I could focus on my work without ever being bogged down by paperwork.

The reason I can put my work at the forefront is an extremely supportive family. None of this would have been possible without the effort and encouragement of my parents, who have placed me and my career ahead of everything else. Sneha has been my motivation through this final phase of my Ph.D.. My extended family, Shri and Shaila Mate, Prasanna, Madhuri, and Amruta Mate have always been around for me. Games with the badminton group at Caltech were a constant source of refreshment (and exercise!). And finally, my family away from home: Mansi, Prabha, Pinkesh, Jayakrishnan, Sushree, Shriharsh, and Ishwari, who have done everything possible to keep me happy. Thank you. *You're awesome*!

Abstract

My thesis centers around the study of neutron stars, especially those in massive binary systems. To this end, it has two distinct components: the observational study of neutron stars in massive binaries with a goal of measuring neutron star masses and participation in NuSTAR, the first imaging hard X-ray mission, one that is extremely well suited to the study of massive binaries and compact objects in our Galaxy.

The Nuclear Spectroscopic Telescope Array (NuSTAR) is a NASA Small Explorer mission that will carry the first focusing high energy X-ray telescope to orbit. NuSTAR has an order-of-magnitude better angular resolution and has two orders of magnitude higher sensitivity than any currently orbiting hard X-ray telescope. I worked to develop, calibrate, and test CdZnTe detectors for NuSTAR. I describe the CdZnTe detectors in comprehensive detail here—from readout procedures to data analysis. Detailed calibration of detectors is necessary for analyzing astrophysical source data obtained by the NuSTAR. I discuss the design and implementation of an automated setup for calibrating flight detectors, followed by calibration procedures and results.

Neutron stars are an excellent probe of fundamental physics. The maximum mass of a neutron star can put stringent constraints on the equation of state of matter at extreme pressures and densities. From an astrophysical perspective, there are several open questions in our understanding of neutron stars. What are the birth masses of neutron stars? How do they change in binary evolution? Are there multiple mechanisms for the formation of neutron stars? Measuring masses of neutron stars helps answer these questions. Neutron stars in high-mass X-ray binaries have masses close to their birth mass, providing an opportunity to disentangle the role of "nature" and "nurture" in the observed mass distributions. In 2006, masses had been measured for only six such objects, but this small sample showed the greatest diversity in masses among all classes of neutron star binaries. Intrigued by this diversity—which points to diverse birth masses—we undertook a systematic survey to measure the masses of neutron stars in nine high-mass X-ray binaries. In this thesis, I present results from this ongoing project.

While neutron stars formed the primary focus of my work, I also explored other topics in compact objects. appendix A describes the discovery and complete characterization of a 1RXS J173006.4+033813, a polar cataclysmic variable. appendix B describes the discovery of a diamond planet orbiting a millisecond pulsar, and our search for its optical counterpart.

Contents

List of Figures

List of Tables

List of Acronyms

ADC: analog-to-digital convertor
AGN: active galactic nuclei
ASIC: application-specific integrated circuit
BH: black hole
cps: counts per second
DBSP: double beam spectrograph
EOS: equation of state
ESD: electrostatic discharge
FITS: flexible image transport system
FOV: field of view
FPGA: field programmable gate array
FPM: focal plane module
FWHM: Full Width at Half Maximum
GUI: graphical user interface
HEASARC: high energy astrophysics science archive research center
HMXB: high-mass X-ray binary
HPD: half power diameter
HV: high voltage
IC: integrated circuit
IR: infrared
LRIS: low resolution imaging spectrograph
MISC: minimum instruction set computer
MOC: mission operations center
NRL: naval research laboratory
NS: neutron star
NuSTAR nuclear spectroscopic telescope array
NuSTARDAS: NuSTAR data analysis software
OGIP: office of guest investigator programs
PH: pulse height
PI: pulse invariant
PMT: photomultiplier tube
PSF: point-spread function
RMF: redistribution matrix file
QE: quantum efficiency
RV: radial velocity
SAA: south atlantic anamoly

SFXT: supergiant fast X-ray transient
SMBH: super massive black hole
SMEX: small explorer
SNR: signal-to-noise ratio
SOC: science operations center
SRL: space radiation laboratory
ToO: target of opportunity
XRB: X-ray binary

Part One: NuSTAR

Characterization of
CdZnTe Detectors for NuSTAR

Image: NuSTAR satellite at Orbital Sciences Corporation

Chapter 1

NuSTAR: The Nuclear Spectroscopic Telescope Array

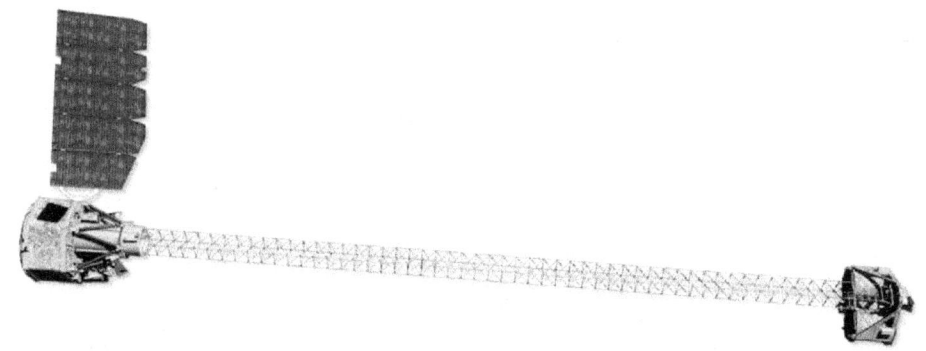

1.1 Hard X-rays

As the universe evolves around us, astronomers examine it from a single vantage point to piece together its working. The pieces of this jigsaw puzzle are information obtained from various wavelengths. Evolving well beyond the old optical telescopes, observational astrophysics now assimilates information from the entire electromagnetic spectrum: extending from the low energy radio waves to high energy gamma rays, spanning over 10 orders of magnitude in energy. Astrophysics also incorporates information from non-electromagnetic channels like cosmic rays, neutrinos and eventually will directly utilize gravitational waves. Harwit (2003) makes an excellent case that technological breakthroughs have ushered in major advances in our knowledge of the cosmos: sometimes by opening up a new energy range for study, at other times by greatly increasing the sensitivity in an energy range.

A relatively underexplored part of the electromagnetic spectrum is the hard X-ray band. X-rays are high energy photons, with energies from few hundred electron volts (eV) to several hundred kilo electron volts (keV) (figure 1.1). This interval is loosely divided into soft X-rays and hard X-rays. Here, we will refer to the energy range covered by NuSTAR (6 to 80 keV) as the Hard X-ray band.

Extending sensitivity to high energy X-rays enables studies of nonthermal processes, often masked at lower energies by thermal plasma emission. For example, synchrotron radiation from hot ionized gas in galaxy clusters is detected in radio wavelengths. However, neither the magnetic field strength nor the number density of electrons can be independently calculated from these data (Rephaeli et al., 2008).

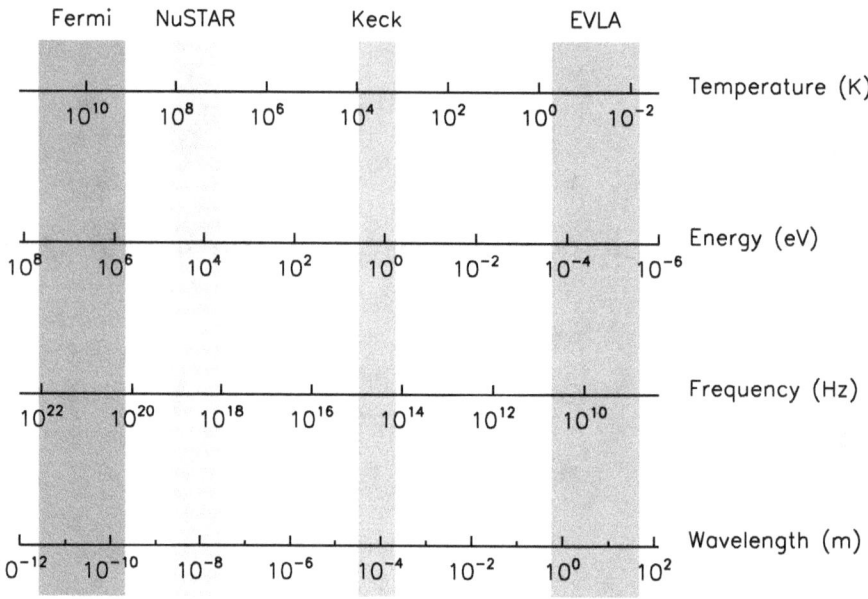

Figure 1.1. The electromagnetic spectrum. The four axes from bottom to top show the wavelength (m), frequency (Hz), energy (eV), and a characteristic temperature (K) for radiation ranging from high energy gamma rays on the left to radio waves on the right. Shaded areas represent energy ranges for a few telescopes. NuSTAR will probe the hard X-ray band from 6 keV to 80 keV ($0.15 - 2$ Å).

The same electrons responsible for the synchrotron radiation will Compton upscatter cosmic microwave background (CMB) photons into the X-ray band. In the soft X-ray band, this radiation is dwarfed by thermal emission from the cluster. Hard X-ray observations are the key to detecting and characterizing the nonthermal emission to further our understanding of the intracluster medium (ICM).

Enhanced sensitivity at higher energies also opens unique opportunities to study emission from radioactive decay. For example, both core collapse and thermonuclear supernovae produce radioactive ^{44}Ti, which emits nuclear deexcitation lines at 68 and 78 keV during radioactive decay. Theoretical predictions for ^{44}Ti production in both types supernovae are widely scattered, and are sensitive to the turbulence and asymmetry in the explosion. For core collapse supernovae, the amount of ^{44}Ti produced depends on the location of the mass cut: the boundary ejecta and material that falls back. The total luminosity in these lines provides a direct measure of the amount of ^{44}Ti produced, thus constraining the explosion physics.

Hard X-ray radiation is highly penetrating. Some sources, e.g., Compton thick AGN, are enshrouded in dust and gas, which absorb and reprocess most of the source radiation. As the absorption cross section for X-rays in material of typical abundances decreases sharply with energy, these sources can still be studied in hard X-rays.

Sensitive hard X-ray telescopes can provide valuable insights into the physics of these and many other astrophysical sources. I discuss current hard X-ray telescopes in Section 1.2, followed by an overview of a next-generation hard X-ray telescope, NuSTAR (Section 1.3).

1.2   Hard X-ray Telescopes

As the Earth's atmosphere is opaque to X-rays, developments in X-ray astronomy lagged significantly behind those in optical or radio astronomy. In 1948, a set of detectors on board a V2 rocket detected X-rays from the Sun—the first detection of non–terrestrial X-rays (Tousey et al., 1951). It took well over a decade for the detection of a cosmic point source: a feat accomplished by Giacconi et al. (1962) by the discovery of Sco X-1. Ever since, the history of X-ray astronomy has been marked by technological breakthroughs going hand-in-hand with astronomical motivation to make better and better satellites,[1,2] which are many orders of magnitude more sensitive than these pioneering experiments (Table 1.1).

These developments in technology warrant a bit more attention to place NuSTAR in context. The most basic elements of an X-ray telescope are the light gathering aperture and the detectors. Early X-ray experiments had very little directionality. Solar X-rays were first discovered by a crude pinhole camera payload on board a Naval Research Laboratory rocket, with no means to collimate or focus the incoming radiation. The next evolutionary step was to add collimators in front of the detectors. Collimators enable basic imaging of the sky in the same way a single dish radio telescope does: by pointing and imaging one field of view at a time. They are still used primarily for X-ray timing, for example in the Proportional Counter Array on board the Rossi X-ray Timing Explorer (RXTE; Jahoda et al., 1996). Finer imaging

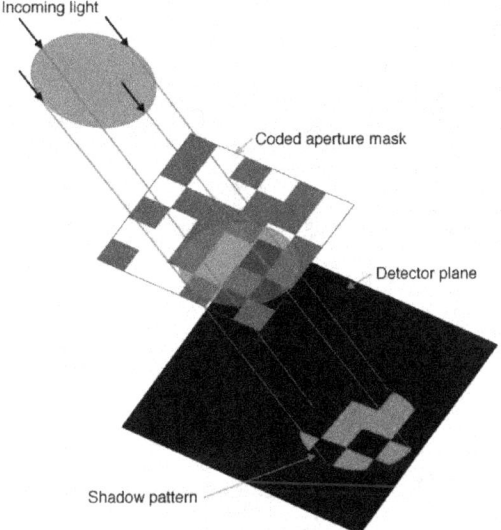

Figure 1.2. Imaging with a coded aperture mask. When light from astrophysical sources (red) is incident on the coded aperture mask (blue), it casts a particular shadow on the detector plane (black). The shape of the shadow and its location on the detector uniquely determine the location of the source in the sky. If the source extent is greater than the angle subtended by an individual mask "pixel" on the detector, then source morphology can be inferred from the image as well.
Image from http://swift.sonoma.edu/about_swift/instruments/bat.html.

---

[1] https://en.wikipedia.org/wiki/History_of_X-ray_astronomy
[2] http://heasarc.gsfc.nasa.gov/docs/heasarc/headates/heahistory.html

Table 1.1.　Currently operating X-ray telescopes compared to NuSTAR

Telescope / Instrument	Detector type	Energy range (keV)	Energy resolution (keV)[a]	Optics type	Effective area (cm^2)	Angular resolution (FWHM)	Field of view	Launch
Chandra ACIS	CCD	0.1–10	0.13	Focusing	235	1″	17′	1999
XMM-Newton EPIC PN	CCD	0.2–12	0.13	Focusing	851	~ 6″	30′	1999
Suzaku XIS	CCD	0.2–12	0.12	Focusing	1000	< 1.5′	19′	2005
Suzaku HXD	PIN diodes	10–60	3	Collimators	140	34′	34′	2005
Integral IBIS/ISGRI	CdTe	15–1000	9% (100 keV)	Coded Aperture	~ 2600 cm^2	12′	19° [b]	2002
Swift BAT	CZT	15–150	7	Coded Aperture	5240	17′	1.4 sr [c]	2002
NuSTAR	CZT	6–80	1	Focusing	920	10″	13′	2012

Note. — Data sources: Suzaku PIN: Takahashi et al. (2007), Integral: Ubertini et al. (2003), Swift: Barthelmy et al. (2005), NuSTAR: Harrison et al. (2010), all others from http://heasarc.nasa.gov/docs/heasarc/missions/comparison.html, retrieved on 2012 March 12. Other relevant high energy telescopes not mentioned here include HETE-2, MAXI, AGILE, Fermi and the upcoming HXMT, SRG (with ART-XC), Astro-H, AstroSat and GEMS.

[a]Energy resolution at 6 keV for soft X-ray instruments and 60 keV for hard X-ray instruments.

[b]Partially coded FOV. Fully coded field is 9°.

[c]Half coded field. 1.4 sr = 4600 sq. deg.

is achieved by using coded aperture masks—which uses the principle of pinhole cameras. The aperture consists of a mask with a non–repeating pattern of holes, placed some distance away from a detector (Figure 1.2). The location of a point source in the sky can be inferred from the shadow it casts. The sky image is reconstructed by deconvolving the observed image with the pattern of holes in the aperture. As in pinhole cameras, the angular resolution is governed by the size of holes in the mask and the mask-to-detector distance. Coded aperture masks have a wide field of view, making them the preferred instruments for monitoring transients. Current coded aperture mask instruments include the Burst Alert Telescope (BAT; Barthelmy et al., 2005) on *Swift* as well as IBIS (Imager on Board the *Integral* Spacecraft) and ISGRI (the *Integral* Soft Gamma-Ray Imager) on *Integral* (Ubertini et al., 2003; Lebrun et al., 2003).

A major breakthrough in soft X-ray astronomy came with the development of focusing optics. Unlike visible light, X-rays do not reflect near normal incidence. The index of refraction of solids for X-rays is slightly lower than unity. Hence, if X-rays are incident on a surface at incidence (or graze angles) below the critical angle, they undergo total external reflection. Based on this principle, the Einstein Observatory (HEAO 2; Giacconi et al., 1979) became the first satellite to use focusing X-ray optics. Today, focusing optics are regularly employed in soft X-ray telescopes. Most notable among them is *Chandra*, which attains subarcsecond angular resolution (Weisskopf et al., 2000).

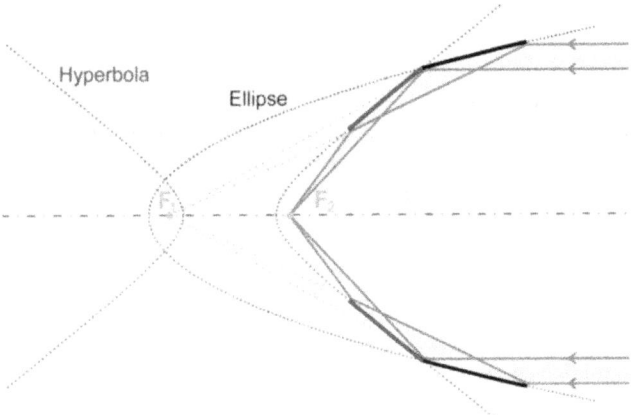

Figure 1.3. A sketch of Wolter-I focusing optics (Wolter, 1952a,b). X-rays incident from the right side are first reflected toward the focus F_1 by a parabolic surface, shown in solid black. These rays undergo a second reflection from a hyperbolic surface (shown in solid blue) to converge at the focus F_2. The two-mirror arrangement reduces the total focal length by roughly a factor of two, and improves image quality across the entire field of view.
Image reproduced with permission from http://www.x-ray-optics.de, retrieved 2012 March 07.

Focusing telescopes concentrate light from the source onto a detector much smaller than the telescope aperture. Since sources can be extracted from smaller parts of the detector, the contributions from astrophysical and detector backgrounds are greatly reduced relative to a coded aperture mask. For example, NuSTAR has a collecting area of $\sim$ 400 cm^2 at 20 keV. Hard X-rays from a point source are focused into a few square-millimeter spot. As compared to a coded aperture mask where the detector area is about a factor of two larger than the aperture, this reduces the background in the extraction region by 10^4 and improves the SNR by a factor of 100 in background–limited observations. Compared to coded aperture masks, focusing telescopes are also more sensitive to diffuse sources.

In spite of the advantages they offer, several technical challenges have prevented adaption of focusing optics to hard X-ray. The critical angle below which X-rays undergo total external reflection varies approximately linearly with wavelength. Hence, for a given focal length, the maximum aperture area of a hard X-ray telescope working at 10 keV that relies only on critical angle reflection is approximately one-hundredth that of a soft X-ray telescope working at 1 keV.

One way to overcome the critical angle limitation is to use multilayer coatings (Christensen et al., 1992; Madsen et al., 2009; Christensen et al., 2011). A multilayer is a stack of thin, alternating layers of two different materials. Multilayers extend reflectivity beyond the critical angle using Bragg reflection. Because partial reflection happens at interfaces between materials of different index of refraction, the stack acts as a periodic lattice, enhancing reflectivity by constructive interference where the Bragg condition is satisfied: $2d \sin \theta = m\lambda$. Here, d is the bilayer thickness, θ is the angle of incidence, λ is the wavelength of the incident photon, and m is the order of reflection. A fixed-thickness multilayer design will lead to enhanced reflectivity at fixed energies where the Bragg condition is satisfied. Broadband reflectivity can be enhanced by using depth-graded multilayers, where a varying thickness in the d-spacing as a function of depth in the coating shifts the Bragg peaks through the spectrum. The reflectance at an interface is proportional to the density (or refractive index) contrast between the two materials. Hence, multilayers are composed of a high density (high refractive index) and a low density material, like tungsten and silicon. The effect of the multilayers is to increase the reflectance angle above the critical angle, with reflectance depending on the minimum d-spacing and design of the layers (Mao et al., 1999). Even with multilayers, extending reflectivity to high energies required telescopes with graze angles smaller than utilized in soft X-rays. For a single surface, the ratio of mirror surface area to the projected collecting area is very low. Collecting area is increased by nesting multiple optics with the same focal point. Hence, many mirrors with large surface area need to be coated with multilayers and aligned within strict tolerances: a technically challenging procedure. Developing depth graded multilayer coated optics was a key item enabling the NuSTAR telescope.

Of equal importance to enabling NuSTAR was developing hard X-ray position sensitive detectors. X-ray detectors employ a wide range of technologies ranging from gas proportional counters (e.g., in the upcoming *AstroSat*; Agrawal, 2006) to fine pixelled CCD detectors in focusing soft X-ray telescopes. However, these detectors have certain limitations (Section 2.1). Proportional Counter Arrays have to strike a trade-off between bulkiness and efficiency. Silicon detectors like CCDs are relatively transparent in the hard X-ray band. Some Coded Aperture Mask instruments use Cadmium Zinc Telluride (`CdZnTe`) detectors with large pixels for imaging. NuSTAR and other upcoming focusing hard X-ray telescopes require detectors with a pixel pitch less than a millimeter, without compromising on quantum efficiency or energy resolution. The development and calibration of such `CdZnTe` detectors at Caltech is the focus of the first half of this thesis.

1.3 NuSTAR

The Nuclear Spectroscopic Telescope Array (NuSTAR) is a NASA small explorer mission that will carry the first focusing hard X-ray optics into space (Harrison et al., 2010). With a pair of coaligned focusing telescopes (Figure 1.4), NuSTAR has an order of magnitude better angular resolution and is two orders of magnitude more sensitive than any existing hard X-ray instrument (Table 1.2, Figure 1.5).

NuSTAR has four primary science goals:

1. Locate massive black holes,

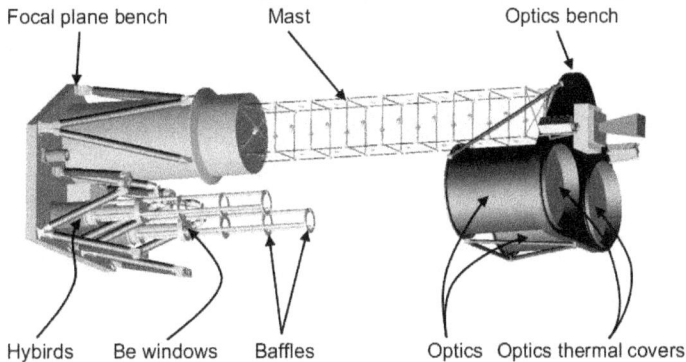

Figure 1.4. A schematic of NuSTAR. The satellite consists of two identical coaligned telescopes. Both optics are mounted on the "optics bench," which is connected to the "focal plane bench" by a deployable mast. The mast is in a stowed configuration at launch, and extends to 10.15 meters when deployed. Detectors (known as hybrids) are housed inside two focal plane modules which contain active shielding, readout electronics, and beryllium entrance windows. The optics are 38 cm in diameter.

Table 1.2.   Sensitivity of hard X-ray telescopes

Satellite (Instrument)	Energy range	Sensitivity
Integral (ISGRI)	20–100 keV	$\sim$ 500 μCrab (>Ms exposures)
Swift (BAT)	15–150 keV	$\sim$ 800 μCrab (>Ms exposures)
NuSTAR	10–30 keV	$\sim$ 0.7 μCrab (1 Ms)

2. Study the population of compact objects in the Galaxy,

3. Understand explosion dynamics and nucleosynthesis in core collapse and Type Ia supernovae,

4. Constrain particle acceleration in relativistic jets in supermassive black holes.

In addition, the science team has identified several other compelling science programs (Table 1.3). Many of these science goals have been incorporated in the observing plan for the two year baseline mission. A Guest Observer (GO) program will be proposed at the end of the baseline mission to broaden the scientific output from the mission.

NuSTAR builds on the technology developed for the balloon mission "HEFT" (High Energy Focusing Telescope; Harrison et al., 2006). NuSTAR's performance is made possible by leveraging three key technological breakthroughs: efficient hard X-ray focusing optics, state-of-the-art CdZnTe detectors, and a deployable mast. Below I describe the telescope design in more detail.

NuSTAR employs low grazing angle focusing optics which are conical approximations to the Wolter-I design (Hailey et al., 2010). Each of the two optics modules on board the spacecraft has 133 concentric, confocal shells with a focal length of 10.15 m. Individual optics shells are made by slumping glass on cylindrical mandrels. These sections are then clamped into place, with precisely machined graphite spacers

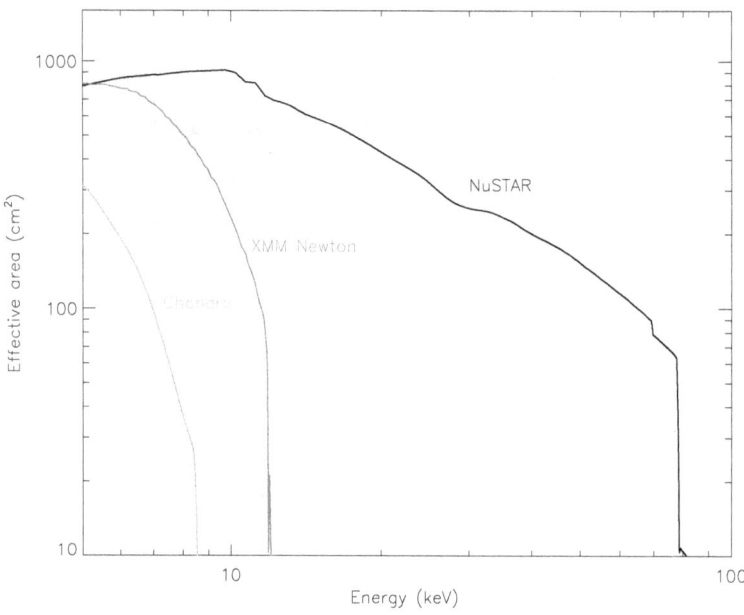

Figure 1.5. Comparing the NuSTAR effective area (both telescopes combined) to *Chandra* and *XMM-Newton* as a function of energy. NuSTAR utilizes a low graze angle design with depth-graded multilayer coatings to extend the sensitivity to 80 keV. The sharp cutoff at 80 keV is caused by a K-shell absorption edge in platinum used in the coatings.

that constrain the glass into the appropriate conical shape. The high energy reflectance of shells is enhanced by using multilayer coatings (Section 1.2). The coatings employ a combination of W/Si bilayers on the outer 43 shells, and Pt/C on the inner 90 shells. Pt has a K-shell absorption edge produces a sharp drop in the effective area at 79 keV.

Table 1.3: A representative list of NuSTAR science targets

Galactic plane survey	Magnetars	AGN physics
Sgr A*	X-ray binaries	Compton-thick AGN
Supernova Ia ToO	Pulsars	Blazars
Supernova remnants	Gamma-ray binaries	Starburst galaxies
Solar physics	Extragalactic surveys	Galaxy clusters
Flaring protostars	Ultraluminous X-ray sources	Radio galaxies
Planetary wind nebulae	Targets of opportunity (ToO)	ULIRGs

The optics have an angular resolution of $\sim$ 12″ (FWHM)[3], and a field of view of $\sim$ 10′. The reflectivity of optics shells starts decreasing with increasing angle of incidence of photons. This effect is

---

[3]The half power diameter (HPD) is $\sim$ 50″.

more pronounced at higher energies and the field of view drops to 6′ at 60 keV (Figure 1.6).

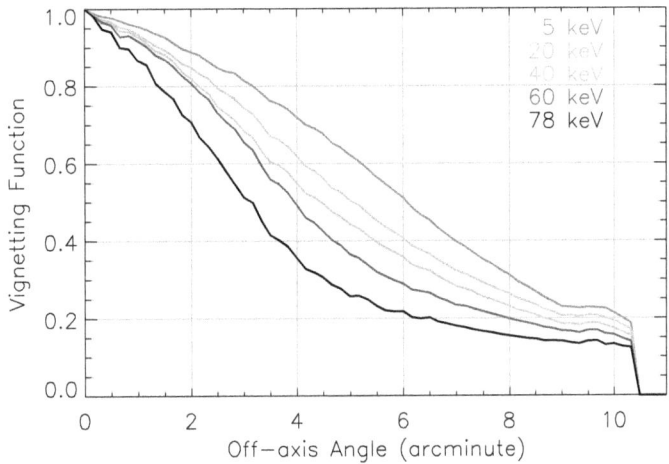

Figure 1.6. Off-axis response of NuSTAR as a function of angle, for various energies. The on-axis response is normalized to unity independently for each energy. The field of view is 10′ in diameter at 10 keV and decreases to about 6′ at 60 keV.

Each telescope has a corresponding Focal Plane Module (FPM) consisting of four 32×32 pixel `CdZnTe` detectors. These detectors have energy resolution of $\sim 1\%$ and high quantum efficiency over the entire NuSTAR energy range. Detectors are discussed in greater detail in Chapter 2.

Small explorer (SMEX) missions are NASA's smallest astrophysics platform and the launch vehicle fairing cannot accommodate a fixed 10 m telescope. To overcome this limitation, the instrument is launched in a compact stowed configuration. After launch, a deployable mast developed by ATK Space Systems, Goleta[4] extends to achieve the 10.15 m focal length.

NuSTAR will be launched in the Summer of 2012 on a *Pegasus XL* rocket by Orbital Sciences Corporation into a 6° inclination, 575×600 km Low Earth orbit. The orbit is selected to avoid passages in the South Atlantic Anomaly (SAA) where the concentration charged particles in the Earth's atmosphere increases, leading to high background noise. The low earth orbit results in frequent earth occultations for several targets. The satellite does not reorient during earth occultation, decreasing the observing efficiency to near 50% for most targets, leaving it closer to 90% for polar targets.

The mission has a nominal lifetime of 2 years, during which it will address primary science goals. This will include a few Target-of-Opportunity (ToO) observations with a response time of < 1 day. There are no consumables on board and the mission life is limited by the ~ 10 year orbit decay timescale. The data will be downlinked to Malindi, Kenya, and transferred to the Mission Operations Center (MOC) at UC Berkeley. The Science Operations Center (SOC) at Caltech validates the data and converts it to FITS format conforming to OGIP standards. NuSTAR science data have no proprietary period, after a six-month interval to calibrate the instrument and verify performance, all data will be uploaded to the HEASARC[5] public archive within two months of completion of an observation.

---

[4]http://www.atk.com/capabilities_multiple/goleta.asp.

[5]High Energy Astrophysics Science Archive Research Center; http://heasarc.nasa.gov/.

Chapter 2

Cadmium Zinc Telluride Detectors

I discuss the desired properties of hard X-ray detectors in Section 2.1, followed by specifics of NuSTAR CdZnTe detectors (Section 2.2). Section 2.5 deals with details of how a photon is processed and read out by NuSTAR hardware. The rest of the chapter discusses various aspects of data analysis: calculating event rates (Section 2.4), calculating photon energies (Section 2.5), and photon pileup (Section 2.6).

2.1 Hard X-ray Detectors

The performance requirements of detectors are driven by the scientific goals of a mission and the accessible technology. As NuSTAR is an imaging telescope, the primary requirement is to develop imaging detectors with a pixel pitch that optimally samples the point spread function (PSF) of the optics. The FWHM of the optics PSF is $\sim 12''$. At a focal length of 10 m, that translates to a physical size of ~ 600 μm. This requirement drives the pixel size of the NuSTAR detectors. We selected CdZnTe as the material of choice for NuSTAR detectors, to make compact, segmented detectors with good energy resolution. CdZnTe functions well at $0°$–$10°$ C, simplifying the readout electronics. These temperatures can be attained in orbit by passive cooling, thereby eliminating the need for cryogenics.

In hard X-ray, the dominant process by which photons interact with high-Z matter used in X-ray detectors is photoelectric absorption (Longair, 1992, Chapter 4). A photon of energy $h\nu$ can eject electrons with binding energies $E_i \leq h\nu$ from atoms. The energy levels in atoms for which $h\nu = E_i$ are called absorption edges, where the absorption probability of a photon increases sharply as it can interact with electrons from this energy level. For example, Xe has an absorption edge at 34.5 keV (Figure 2.1). For photons with energies greater than the absorption edge, the cross section for photoelectric absorption from this level decreases roughly as ν^{-3}. At a given photon energy, the photoelectric absorption cross section also strongly depends on the atomic number (Z) of the material. At hard X-ray energies, the absorption cross section for K-shell electrons of atoms is proportional to the fifth power of the atomic number. Thus, CdZnTe (mean $Z = 49.1$[1]) absorbs hard X-rays above 10 keV better than, say, Si ($Z = 14$).

In a semiconductor detector, when a photon with energy $h\nu$ ejects an electron with binding energy E_i, the remaining energy $h\nu - E_i$ is converted into kinetic energy of the ejected electron. Part of this energy is lost to the lattice of the substrate, but much of the rest of it serves to excite electrons from the valance band to the conduction band, creating thousands of electron–hole (e^-–h^+) pairs. The total number of e^-–h^+ pairs generated depends on the energy of the incoming photon and the characteristics of the substrate, in

---

[1]Atomic numbers are Cd = 48, Zn = 30, Te = 52.

13

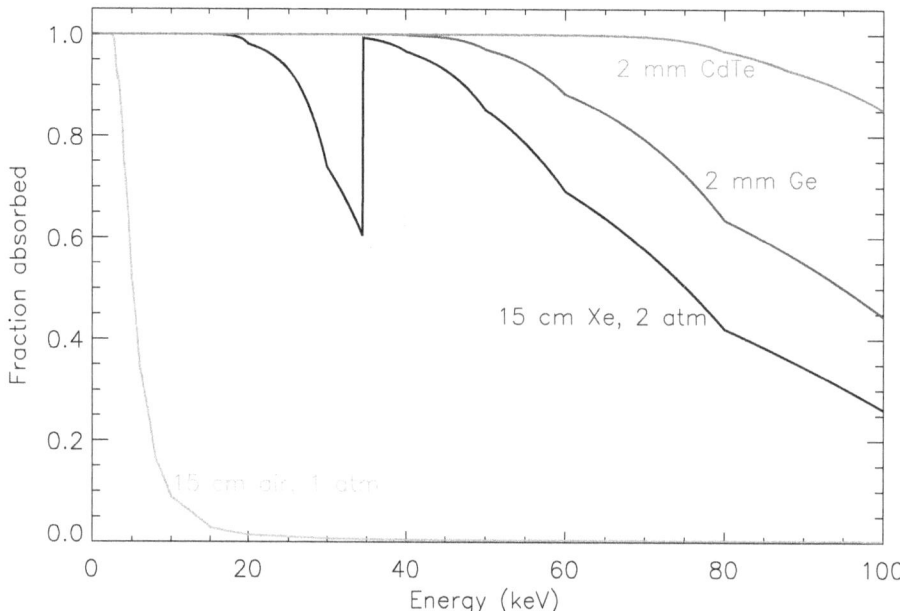

Figure 2.1. Comparing the X-ray absorption properties of various elements. The green curve shows the absorption
    by 15 cm of dry air at 1 atmosphere—this small air column is essentially transparent to X-rays above
    10 keV. The purple curve is the absorption curve for 15 cm of Xe, similar to the gas-filled detector
    LAXPC on the upcoming Astrosat (Agrawal, 2006). The opacity rises sharply at the 34.5 keV K-edge
    for Xe. Solid state detectors like Ge (blue) perform better. A 2 mm thick CdZnTe crystal like ones used
    in NuSTAR hybrids absorbs almost all X-rays up to 80 keV (red curve, top).

particular the band gap. Thus, we can calculate the energy of the incident photon by measuring the charge
generated in the detector. This process of "event reconstruction" is discussed in more detail in Section 2.5.

Finally, another important consideration for X-ray instruments is the timing capabilities. The times of
arrival of photons carry information about the emission characteristics of the source. For example, pulsations and periodicities in X-ray binaries can be analyzed to understand the properties of the compact object
and orbital characteristics of the system. Typical count rates from astrophysical sources in the hard X-ray
band are rather low: an X-ray source with a flux of about 1 mCrab gives about 10^{-4} photons cm^{-2} s^{-1}
in the 15–50 keV band. With typical apertures of a few hundred square centimeters, the total count rates
remain well within the processing realm of modern electronics.

In summary, a good hard X-ray detector should accurately measure the position, energy, and time of
arrival for every X-ray photon that strikes it, while having high Quantum Efficiency (QE) and space-suitable
features like compact geometry. Requirements for NuSTAR detectors are given in Table 2.1.

Table 2.1. NuSTAR focal plane configuration summary

Parameter	Value
Pixel size	0.6 mm/12.3″
Focal plane size	$13' \times 13'$
Pixel format	32×32
Threshold	2.5 keV (each pixel)
Max processing rate	400 evt/s
Max flux meas. rate	10^4/s
Time resolution	$2\mu s$
dead-time fraction (weak source)	2%

2.2 NuSTAR Detector Architecture

The Space Radiation Laboratory (SRL) at Caltech started developing hard X-ray detectors over a decade ago, for a balloon experiment named "High Energy Focusing Telescope" (HEFT; Harrison et al., 2006). The technologies developed for that first generation of detectors were adapted and refined to make NuSTAR detectors. NuSTAR has two focal plane modules, each consisting of four "hybrids" consisting of a CdZnTe crystal mounted on a custom integrated circuits (Figure 2.2). Each hybrid is an array of 32×32 pixels of 605 μm each. These sizes were determined from considerations of the smallest pixel size which can still incorporate the necessary circuitry, and the largest practical size of a uniform CdZnTe crystal. Hybrids assembled and selected for NuSTAR are designated by numbers, like H82, H95, etc. Let us look at the two components of the hybrid to understand their basic working principles. For more details, see Iniewski (2010), Chapter 3.

The readout circuitry of the hybrid is a custom ASIC (Application Specific Integrated Circuit) developed at Caltech and manufactured by ON semiconductors. The ASIC fulfills the requirements for NuSTAR, with low power consumption, low readout noise, and a "rad hard by design" implementation to tolerate the space radiation environment. Each pixel contains a charge sensitive low-noise preamplifier, 16 sampling capacitors, a shaping amplifier, a discriminator, and a latch. The design also incorporate test probes like a programmable analog output for studying with oscilloscopes. A precision test pulsar in each pixel can be used to test the performance of the preamplifiers and measure linearity, offset, and noise characteristics. For NuSTAR, the ASIC is operated in a "charge pump mode," achieving very low electronic noise ($\sim$ 200 eV FWHM). In this mode, it can handle detector leakage currents up to 200 pA. The total noise of the electronics depends on the impedance in the feedback loop, where the capacitance is a dominant term. The detector crystal is directly bonded onto the ASIC, eliminating coupling networks and minimizing the input capacitance. To ensure that the input capacitance at the anode pads is lower than the parasitic capacitance between anode pixel pads (about 300 fF), the ASIC–crystal spacing needs to be $\sim$ 50 μm. This spacing is achieved by bonding a gold wire to the ASIC input pad, and attaching the other side to the CdZnTe using a conductive epoxy. The ASIC is controlled by an external microprocessor. In NuSTAR, a MISC (Minimum Instruction Set Computer) and some peripheral state machines on an Actel FPGA (Field Programmable Gate Array) control the four ASICs in a Focal Plane Module.

The detector substrate is a 2 mm thick CdZnTe crystal, manufactured by eV microelectronics (now

Figure 2.2. A NuSTAR hybrid (shiny square) mounted on a motherboard (green). The hybrid measures about two
centimeters on a side. The top surface is the Pt cathode. A tiny wire from the capacitor "A" forms the
connection for applying a negative high voltage. The three empty quadrants can hold one hybrid each.

EI Detection & Imaging Systems[2]). The front surface seen in Figure 2.2 is a 1000 Å thick continuous
Pt cathode. The anode contacts are a combination of about 1000 Å Pt and 3000 Å gold, laid out in a
605 μm pitch square grid matching the ASIC. Adjacent anode contacts are separated by 50 μm gaps.
Unlike the familiar CCD (Charge Coupled Device) technology, there is no physical insulation within the
semiconductor substrate between pixels. The segmented anode pattern forms the pixels in the detector.
Having a larger spacing decreases the parasitic inter pixel capacitance, and decreases noise—but leads to
higher charge loss (Bolotnikov et al., 1999). Hence, we use the minimum possible gap dictated by EI's
fabrication process. Near the edge of the detector, a guard ring with a tunable bias voltage steers the
electric fields to the edge pixels. The detectors have two "critical edges," where the pattern is laid out
close to the physical edge of the crystal with strict tolerance needs (Figure 2.3). This allows 4 hybrids to
be placed together in a NuSTAR focal plane module with an effective gap less than a pixel. In normal
operation, the cathode is biased at about −300 to −500 V with respect to the anode, using an external
high voltage power supply.

---

[2]http://www.evmicroelectronics.com/

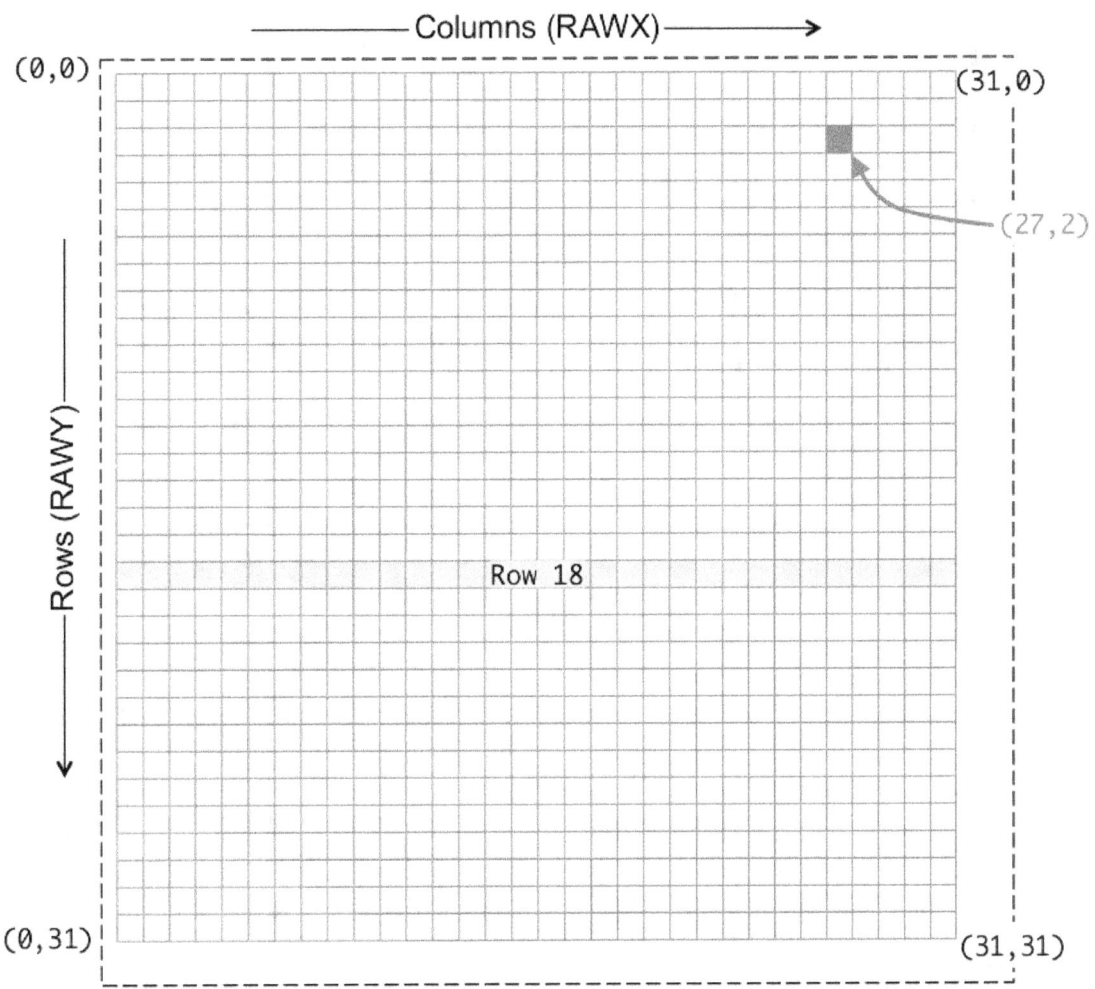

Figure 2.3. Schematic of a NuSTAR `CdZnTe` detector as seen from the top. The gray grid of squares mark the pixels, and the black dashed box is the physical boundary of the detector. The top and left edges are the "critical edges," intersecting at the critical corner. The four hybrids in a FPM are placed with their critical corners at the center. Row numbers increase from from 0 at the top to 31 at the bottom, and are also referred to as `RAWY`. The row with `RAWY` = 18 is highlighted as an example. Column (`RAWX`) numbers increase from left to right. Pixels are numbered as (`COL`, `ROW`) or (`RAWX`, `RAWY`)—for example the pixel highlighted in red is (27, 2).

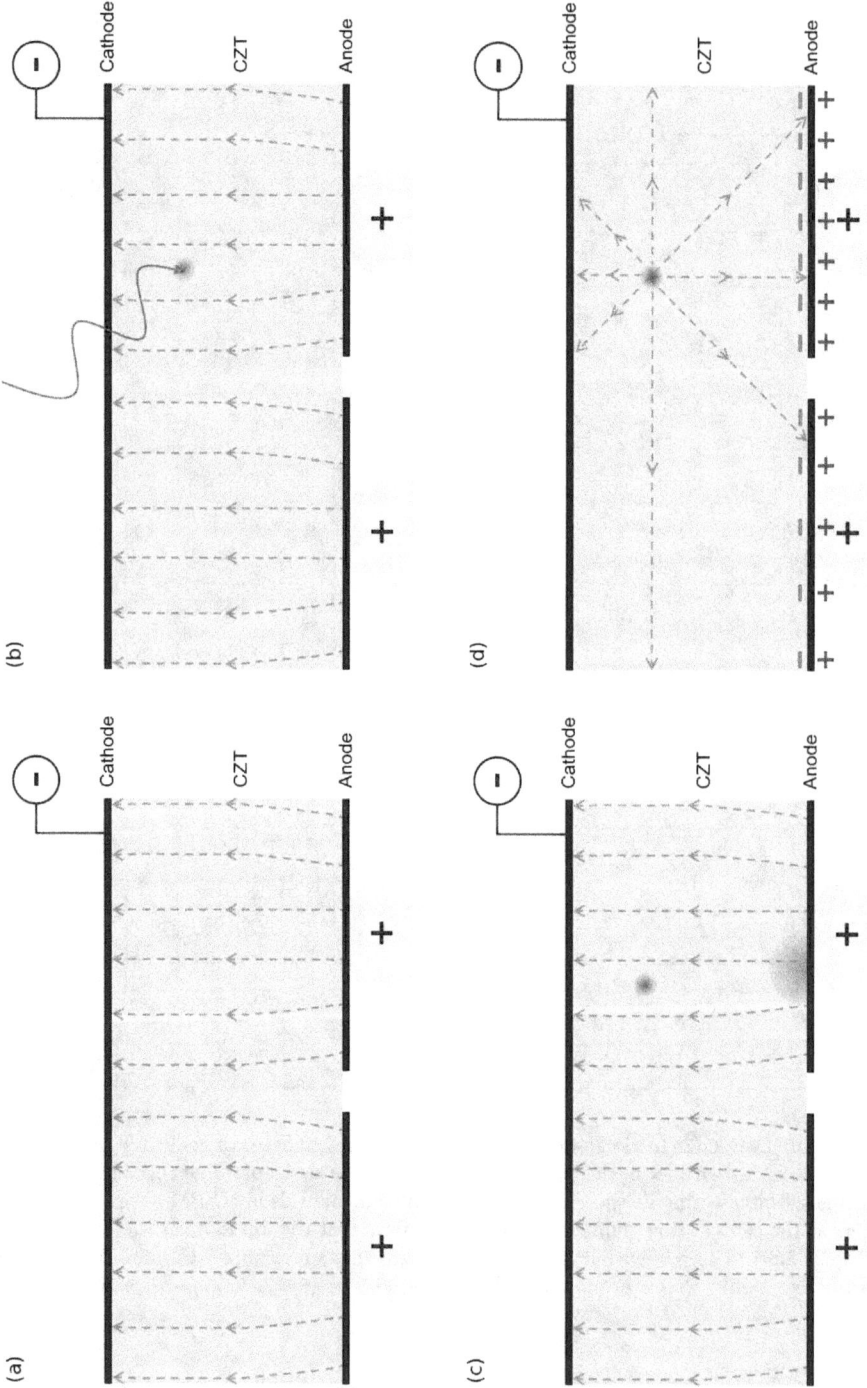

Figure 2.4. Working principle of CdZnTe detectors. (a) The NuSTAR hybrids consist of a CdZnTe crystal (shown in gray) with a continuous Pt cathode (blue, top), biased at -300 to -500 V with respect to a pixelled anode (blue, bottom). The anode contacts are shown in yellow. (b) An incident hard X-ray photon interacts to generate a charge cloud of $e^- - h^+$ pairs. (c) The electrons (shown in red) drift to the anode under the applied potential. The cloud expands due to diffusion and self-repulsion. Holes (light blue) remain frozen in-place. (d) Electric field created by the trapped holes, and the resulting induced charge on the electrodes. For simplicity, the electric field due to the applied voltage is not shown.

2.3   Photon Trigger and Readout

To understand the working of the detector, let us consider a hard X-ray photon incident on the detector (Figure 2.4). It preferentially interacts with an inner-shell electron, eventually creating e^-–h^+ pairs as discussed in Section 2.1. For CdZnTe, the "effective band gap," or energy consumed per e^-–h^+ pair is about 4.85 eV. The electron cloud drifts towards the anode under the applied voltage. As the e^- cloud drifts, it expands due to diffusion and self-repulsion. The electrons follow the electric field and are collected on one or more anode pixel pads. The mobility of holes in CdZnTe is significantly lower than electron mobility, so the holes essentially remain frozen in place at the interaction site. They induce a mirror charge on the anode pixels, with opposite polarity as the electron signal.

A charge–sensitive preamplifier connected to each anode pads produces two outputs. A current output is sequentially routed to a bank of sixteen sampling capacitors in a round-robin fashion. The programmable sampling interval is normally set to $1\mu s$[3]. The second preamplifier output is a voltage signal proportional to the current. This output is fed in to a shaping amplifier with shaping time of approximately 0.5 μs. A discriminator compares the shaping amplifier output to an externally set voltage to generate a hardware trigger. This external voltage thus sets the trigger threshold, viz. the minimum charge deposit required to trigger a pixel. Due to manufacturing tolerances, this threshold corresponds to a slightly different photon energy for each pixel and needs to be measured during calibration.

The hardware trigger for pixel is stored in a latch. These hardware triggers are sequentially combined in OR gates in a row to produce one trigger signal per row (at column 31). These 32 signals are combined in a OR gate into a single ASIC hardware trigger. The trigger outputs from all 4 ASICs are routed off-board to the FPGA, where all or a subset of them can be OR-ed to produce a FPM trigger signal. So far, the trigger processing is asynchronous. It is converted to a synchronous pulse in two clock ticks and fed to a state machine called the "pico processor" on the FPGA, which issues a "lockout" signal to stop processing further triggers in the ASICs. Thus, the "coincidence window" between the hardware trigger and the lockout signal is two to three clock cycles. We will revisit this number in Section 2.4. The round-robin sampling of capacitors is stopped after 6 more clock cycles, so that the capacitor banks have eight pre- and post- trigger samples each.

The MISC controls the processing and readout of the trigger. The event is rejected as a noise event if two hybrids triggered within the coincidence window. If the trigger was from a single hybrid, the MISC reads out a trigger map consisting of one bit per pixel, from the latch of each pixel. This map is then searched for hardware triggers starting from pixel (31, 31), progressing upwards along a column to pixel (31, 0), and repeating the procedure for columns 30–0. From charge sharing considerations, we expect that astrophysical hard X-ray photons will not deposit significant charge in more than four pixels. So, the search for hardware triggers is terminated after finding the first four triggers in this order. The number of processed triggers can be changed by a software setting.

At this point, the sixteen capacitors in each pixel hold 8 charge samples before the trigger and 8 after. We denote these samples by $s_0, s_1 - s_{15}$. In case multiple pixels triggered within the coincidence window, the sampling process stops after 8 clock ticks from the *first* hardware trigger. These capacitors are read out one by one to a charge-rebalance Analog-to-Digital Converter (ADC). To first order, the relation between the value read out and deposited charge is independent of the actual on-chip capacitances, making it robust

---

[3]The onboard clock frequency is $f_{clk} = 14.7456$ MHz. Each capacitor is connected for 15 clock ticks, giving $t_{samp} = 15/f_{clk} = 1.017$ μs.

to manufacturing variations. For each pixel, we calculate the pre- and post- trigger charge levels:

$$\text{PRE} \;=\; s_0 + s_1 + s_2 + s_3 + s_4 + s_5, \tag{2.1}$$

$$\text{POST} \;=\; s_9 + s_{10} + s_{11} + s_{12} + s_{13} + s_{14}. \tag{2.2}$$

The charge deposit in each pixel is calculated as the difference between the POST and PRE sums. The pixel with the highest charge deposit is selected as the central trigger pixel. The misc then reads out the eight nearest neighbors of only this pixel and calculates their PHAs. If any of the neighbors had hardware triggers, they have been read out already and are not processed again.

The exact time of the photon trigger within the 1 μs sampling window can be estimated by assuming a linear rise in deposited charge for the first few samples after the trigger. We calculate a weighted mean of these samples as the time-of-rise estimator, which is used later in event reconstruction to correct for the time shifts in the sampling that affect the energy measurement (Section 2.5). The estimator is defined as

$$
\begin{aligned}
\text{TOR} \;&=\; \frac{-3(s_6 - s_5) - 1(s_7 - s_6) + 1(s_8 - s_7) + 3(s_9 - s_8)}{(s_6 - s_5) - (s_7 - s_6) + (s_8 - s_7) + (s_9 - s_8)} \\[4pt]
&=\; \frac{3s_5 - 2s_6 - 2s_7 - 2s_8 + 3s_9}{s_9 - s_5}, \\[4pt]
\text{TOR} \;&=\; \frac{\text{NUMRISE}}{\text{DENRISE}},
\end{aligned}
\tag{2.3}
$$

where the 16 capacitor samples are numbered $s_0 - s_{15}$ as before. The MISC calculates the NUMRISE and DENRISE terms and the division is carried out in post processing.

Each photon event is packaged into an "event" and immediately sent to the Central MISC or central processor on board the NuSTAR satellite. The event data (Figure 2.5) contain all relevant event information, including

- Packet synchronization header.

- Column (RAWX) and row (RAWY) of the central pixel, number of the starting capacitor for the sixteen samples (S_CAP).

- FPM and detector number to identify among the eight hybrids on board NuSTAR.

- Timing information: Time since the once-per-second frame sync signal, live-time since last event, and time since last charge pump reset.[4]

- NUMRISE and DENRISE for the time-of-rise estimator.

- PRE and POST trigger sums for the central pixel and the 8 nearest neighbors.

The event time is referenced to a once-per-second sync signal from the spacecraft. Stability of the onboard clocks allows relative timing accuracy of tens of microseconds. The absolute timing accuracy of NuSTAR is ~ 100 ms, determined by the spacecraft to UTC time synchronization.

The MISC can reject certain types of events based on programmable criteria. Any events with photons detected on two hybrids within the coincidence window are rejected as they likely originate from a particle

---

[4]In charge pump mode, the feedback capacitor on the preamplifier is reset once every millisecond to remove any charge deposited by the leakage current. The pixel is marginally more susceptible to noise for a few microseconds after this reset.

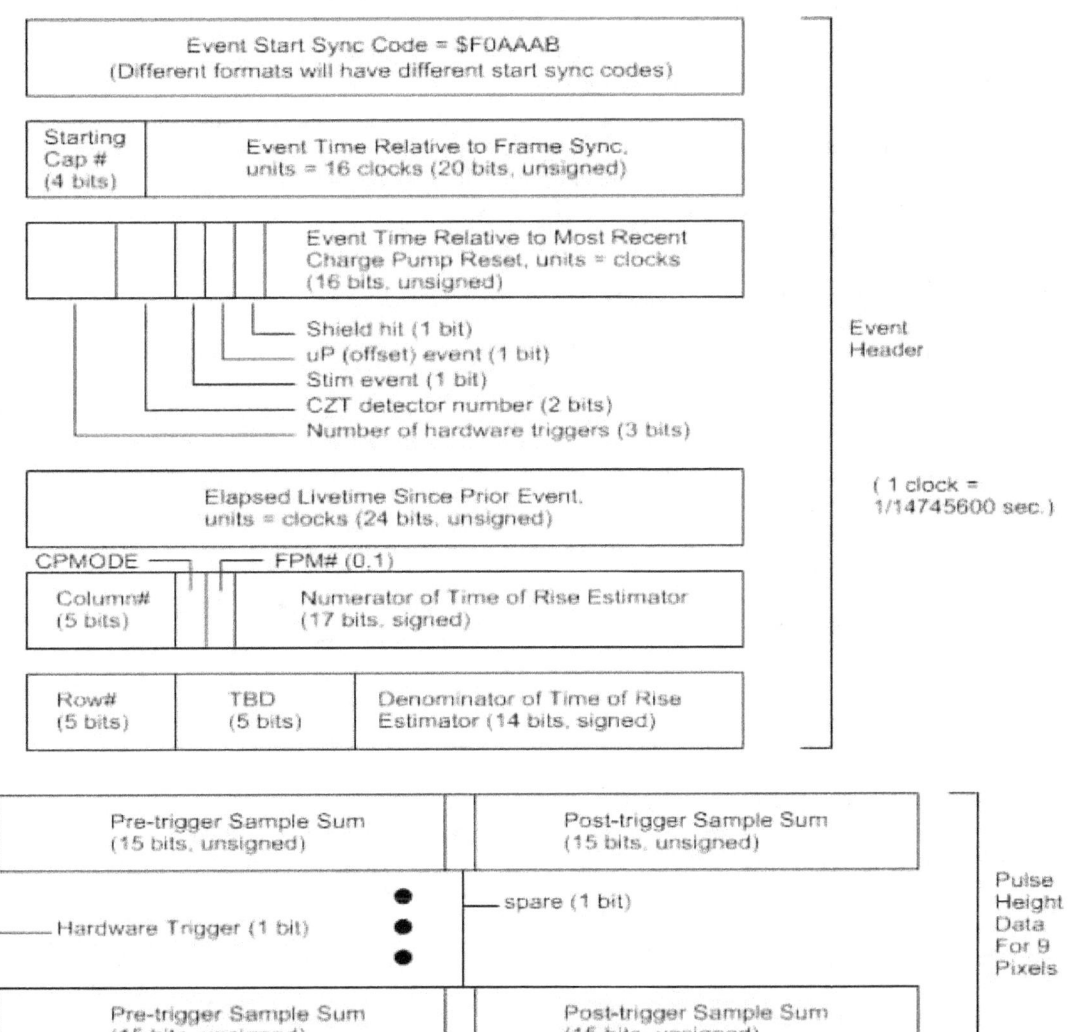

Figure 2.5. NuSTAR event data packet format. Event packets are 56 bytes long and contain all information about a photon event: the PRE and POST trigger sums, time of rise estimator, event timing information, pixel and detector ID, etc. (See Section 2.3 for a detailed description.)

shower induced event. The PMT (Photomultiplier tube) in the active CsI shield generates SHIELDHI and SHIELDLO signals when it detects light above a high and low threshold respectively. A SHIELDHI trigger corresponds to high amplitude signal generated in the PMT. Hence, the anticoincidence window is digitally stretched to 500 μs to allow sufficient time for the PMT to stabilize again before starting event processing again. These signals are digitized and processed by the MISC, where an algorithm computes whether any photon event should be vetoed based on the shield triggers. The MISC software can be set to either discard rejected events or to transmit them in special rejected event packets that contain only relevant information, like time since previous event.

The MISC then resets all latches, initiates the round-robin sampling of capacitors and the detector becomes live again after 8 samples. The entire readout process takes 2.50 ms. Any new photon events during this window are not processed. In other words, the detector is "dead" during this time, an effect discussed in detail in the next section.

2.4   live-time, dead-time, and Event Rates

As discussed in Section 2.3, there is a $\tau_D \approx 2.50$ ms dead-time window in the focal plane after a photon trigger, when that event is being processed. During this window, the focal plane does not trigger on any other incident photons. At low count rates, this does not significantly alter the inferred count rate or spectrum. If the incident rate on a focal plane (4 detectors combined) is high, then it is likely that photons will be incident on the detector during the dead-time windows. These photons are lost as they do not trigger the detector (Figure 2.6).

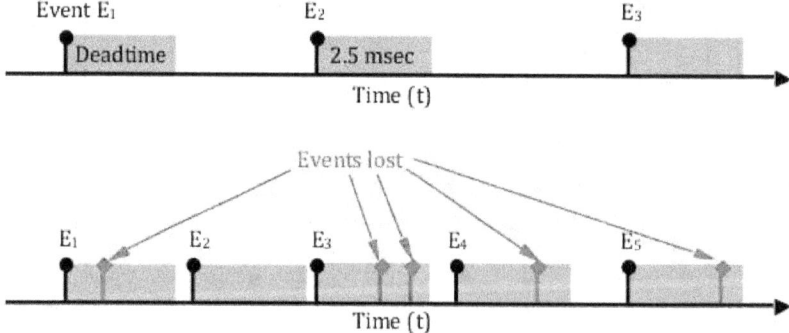

Figure 2.6. Top: Events (E_i) triggering the detector at a low count rate. Gray rectangles denote dead-time windows during which the event is processed. At low count rates, there is a very low probability of receiving photons in the dead-time interval following any trigger. Bottom: At higher count rates, several photons (shown in red diamonds) are incident during the dead-time intervals, and are lost. The measured count rate is thus lower than the incident count rate.

This effect starts becoming significant when the incident count rate R_i becomes high enough that the mean time between events is comparable to the dead-time τ_D. The limiting case is where the detector is triggered as soon as it goes live—this corresponds to $R_{i,\mathrm{crit}} \sim \tau_D^{-1} = 400$ counts s^{-1}.

Since the detector can detect incident photons only in "live" intervals, the incident count rate R_i is given by

$$R_i = \frac{\text{Number of photons}}{\text{Live time}}. \tag{2.4}$$

If we observed N_o photons in a time interval Δt, then the live-time is $t_{\text{live}} = \Delta t - N_o \tau_D$. Substituting this in Equation (2.4), we get

$$
\begin{aligned}
R_i &= \frac{N_o}{\Delta t - N_o \tau_D} \\
&= \frac{N_o/\Delta t}{1 - (N_o/\Delta t)\tau_D} \\
&= \frac{R_o}{1 - R_o \tau_D},
\end{aligned}
\tag{2.5}
$$

where $R_o = N_o/\Delta t$ is the observed (measured) count rate. Equation (2.5) can be inverted to calculate R_o from R_i:

$$R_o = \frac{R_i}{1 + \tau_D R_i}. \tag{2.6}$$

The transformation between R_i and R_o is given in Table 2.2 and plotted in Figure 2.7.

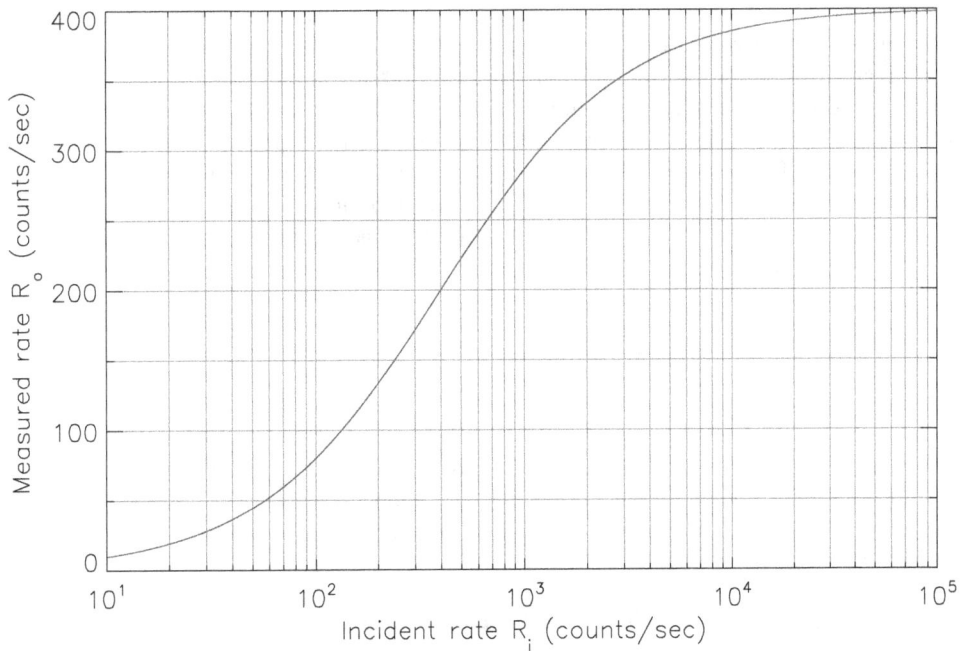

Figure 2.7. Count rate conversion: each event trigger in NuSTAR hybrids is associated with 2.5 ms dead-time. As a result, the measured count rate is a nonlinear function of the incident count rate (Section 2.4, Table 2.2).

Apart from the basic conversion above, there are other effects like rejected events, charge pump resets,

Table 2.2: Observed count rate (R_o) as a function of incident count rate (R_i).

R_o	20	50	75	100	150	180	200	220	240
R_i	21	57	92	133	240	327	400	489	600
R_o	250	260	270	280	290	300	310	320	330
R_i	667	743	831	933	1055	1200	1378	1600	1886
R_o	340	350	360	370	380	385	390	395	399
R_i	2267	2800	3600	4933	7600	10,267	15,600	31,600	159,600

Note. — The incident count rate R_i is for ground calibrations, and does not include effects like shield vetos which are expected in orbit.

and shield veto events which can contribute to detector dead-time. In practice, the count rate is calculated by one of two methods. First, the MISC maintains a live-time counter which, as the name suggests, counts the number of clock ticks for which the detectors were live in a given second. A slight downside to this approach is that only one second averages of live-time are available.

The second method is to calculate the livetime from event data itself. The event data for every photon contains the live-time since last event, known as PRIOR (Figure 2.5). Adding up the PRIORs from all event packets, in principle, gives the total live-time of the detector. However, this is complicated slightly because of rejected events. Consider a case where a photon event is rejected because of multidetector hits or a shield veto. The MISC sends a lockout signal, starts processing the event, and decides to veto it. The detector remains dead for less than the nominal 2.5 ms window. However, this resets the "time since last event" counter. As this event did not enter the data stream, the associated dead-time is not counted when we add up the PRIORs for all accepted events. Rejected event packets, if enabled, contain the live-time prior to each rejected event. Combining this information from accepted and rejected events, we can also calculate the dead-time associated with processing each rejected event. Using all this information we can calculate exact incident count rates.

2.5   Event Reconstruction

The raw spacecraft telemetry packets ("Level 0" data) are converted to produce scientific products ("Level 3") through several stages of processing (Table 2.3). Here, I describe the processing algorithms for calculating the energy of the incident photon using data from the event packets. For analyzing NuSTAR data, these algorithms are implemented in the NuSTAR Data Analysis Software (NuSTARDAS).

The data for each event contains the pre- and post- trigger sums for the trigger pixel and its eight nearest neighbors. First, we calculate the RAWPHAS from these values,

$$\text{RAWPHAS}[9] = \text{POSTPHAS}[9] - \text{PREPHAS}[9]. \tag{2.7}$$

In this discussion, we use the suffix [9] for parameters which are stored separately for each of the nine pixels read out. The relation between the deposited charge and the ADC output varies slightly from capacitor to capacitor. We correct for this effect by subtracting offsets (OFFSET[9]) calculated for the starting capacitor

Table 2.3. Data-processing overview

Stage	Name	Description	Output
1	Data Calibration	Processing of FITS formatted Level 0 telemetry	Level 1a calibrated unfiltered event files
2	Data Screening	Filtering of the calibrated event files by applying conditions on specified attitude/orbital/instrument parameters	Level 2 cleaned event files
3	Products Extraction	Extraction of high-level scientific products (images, light-curves, spectra, exposure maps) from cleaned event files	Spectra, light-curves, images, exposure maps

S_CAP for every pixel.

$$\text{OFFPHAS}[9] = \text{RAWPHAS}[9] - \text{OFFSET}[9] \tag{2.8}$$

We apply a time-of-rise correction to remove the effect of exact time of incidence of the photon within the 1 μs sampling interval. For this, we use the NUMRISE and DENRISE terms calculated by the MISC, and a table of TIMERISE coefficients measured for each pixel:

$$\text{TRPHAS}[9] = \text{OFFPHAS}[9] \times \left(1 + \frac{\text{NUMRISE}}{\text{DENRISE}} \times \text{TIMERISE}[9]\right). \tag{2.9}$$

Next, we apply the "common mode correction" to calculate PHAS. The signal seen by a trigger pixel is the number of electrons collected by that pixel, less the hole charge imaged on that pixel (Figure 2.4). We separate pixels which have charge contributions from the photon (type "$E+$") from type "$E-$" pixels that see only the imaged hole charge. We estimate the hole charge signal from the $E-$ pixels and use it to correct the $E+$ signal. Note that the sign convention here refers to the energy deposited. The actual polarity of the signal is negative because the hybrids read out an electron signal (Section 2.3). The boundary between $E+$ and $E-$ signals, known as the software trigger threshold (EVTTHR), is determined empirically for each pixel. The 9 pixels are divided into three distinct groups: M pixels that are located outside the detector (for edge pixels) or are bad / hot pixels; N pixels (type "$E+$") with $\text{TRPHAS}[9] \geq \text{EVTTHR}[9]$; and $9 - N - M$ pixels (type "$E-$") with $\text{TRPHAS}[9] < \text{EVTTHR}[9]$. The common mode noise term is calculated as

$$\langle E- \rangle = \frac{\Sigma E-}{9 - N - M}. \tag{2.10}$$

We apply this correction only to the "$E+$" pixels, to get

$$\text{PHAS}[9] = \text{TRPHAS}[9] - \langle E- \rangle. \tag{2.11}$$

Values for the $9 - N - M$ type "$E-$" pixels are left unchanged ($\text{PHAS}[9] = \text{TRPHAS}[9]$), while the M outside/bad pixel values are zeroed out ($\text{PHAS}[9] = 0$). Similarly, we also compute SWTRIG[9] for pixels, such that $\text{SWTRIG}[9] = 1$ for the "$E+$" pixels and $\text{SWTRIG}[9] = 0$ for others. We calculate the scalar SURR by adding together the PHAS[9] for pixels with $\text{SWTRIG}[9] = 0$.

This is followed by assigning a GRADE from 0 to 31 to each event, based on the morphology of the SWTRIG[9] map (Figure 2.8). All these steps are implemented in the nucalcpha module of NuSTARDAS.

Next, we convert the PHAS into PIs in three steps. PI (Pulse Invariant) is expressed in units of 40 eV. This number was selected to be a factor of few smaller than the energy resolution of NuSTAR detectors. First, we use the grades to apply pixel- dependent gain and offset corrections to PHAS[9]:

$$\text{PI}'[9] = \text{PHAS}[9] \times \text{GAIN}[9] + \text{OFFSET}[9]. \tag{2.12}$$

The PI$'$ calculation requires interpolation of coefficients in temperature and time. First, the two temperature measurements (in satellite housekeeping data) closest in time to the instant of the photon interaction are identified. For each of these two temperature measurements, a GAIN and OFFSET value is calculated by interpolating between tabulated GAIN, OFFSET values from the two nearest temperatures. The final GAIN[9] and OFFSET[9] values are calculated by interpolating with respect to time between these two coefficients.

The second step is to apply a charge loss correction for event grades 1–8 (two- and three-pixel events), based on the relative amount of charge deposited in the pixels. For grades 1–4, the corrected energy (PI)

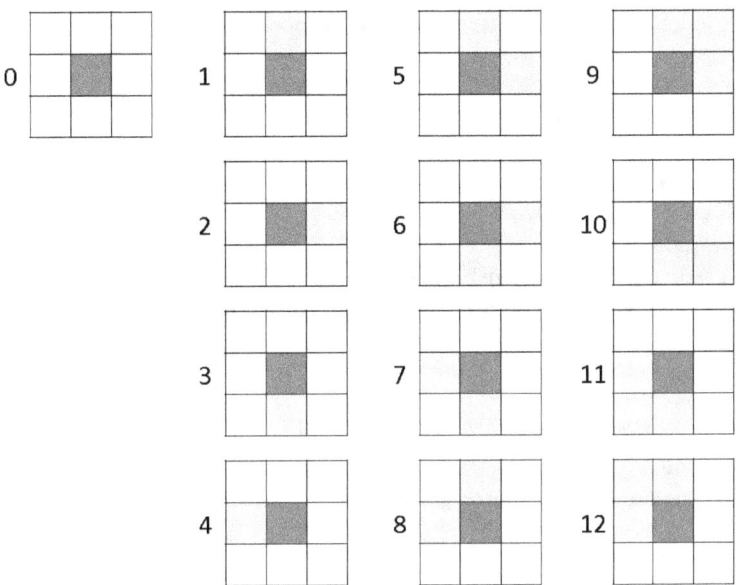

Figure 2.8. Grades are assigned to multiple pixel events based on the morphology of triggered pixels. The pixel grid is oriented as in Figure 2.3. In each case, the central pixel (red) is the one with the highest energy deposit. The nearby pixels shown in yellow have energy deposit higher than the software trigger threshold, and optionally greater than the hardware trigger threshold. Typically, about 90%–95% photon events in NuSTAR hybrids fall under the grades 0–12 shown here.

is given by Equation (2.13):

$$E_{corrected} = rA(\sin\phi + \cos\phi), \tag{2.13}$$

where

$r = \sqrt{E_{center}^2 + E_{second}^2}$,
E_{center} is the grade 0 gain corrected PHA for the center pixel,
E_{second} is the grade 0 gain corrected PHA for the side pixel,
$\phi = \tan^{-1}(E_{second}/E_{center})$,
$A = C_{[X,Y,G]} \times \sin(2\phi)$,
$C_{[X,Y,G]}$ are charge loss correction coefficients for each column (X), row (Y), and grade (G) combination, for grades 1–4.

A fraction of high energy photons cause fluorescence events. In this case, the incident photon interacts in the detector to deposit some energy in a pixel, releasing a Cd or Te florescence photon. This photon may travel to an adjacent pixel and deposit energy in it. Since this is not the same phenomenon as the typical two pixel events, we do not apply a charge loss correction for fluorescence events. So, if $E_{second} = (26.7112 - 1.0) \pm 0.5$ keV (Cd) or $(31.8138 - 1.0) \pm 0.5$ keV (Te) then $E_{corrected} = E_{center} + E_{second}$.

For three pixel events of grades 5–8, the procedure is modified a bit:

$$E_{corrected} = E_{center} + r_1 A_1 \sin \phi_1 + r_2 A_2 \sin \phi_2, \qquad (2.14)$$

where,

$r1 = \sqrt{(E_{center} + E_{third})^2 + E_{second}^2}$,

$r2 = \sqrt{(E_{center} + E_{second})^2 + E_{third}^2}$,

E_{center} is the grade 0 gain corrected PHA for the center pixel,

E_{second} is the grade 0 gain corrected PHA for the pixel with second highest PHA,

E_{third} is the grade 0 gain corrected PHA for the pixel with third highest PHA,

$\phi_1 = \tan^{-1}(E_{second}/(E_{center} + E_{third}))$,

$\phi_2 = \tan^{-1}(E_{third}/(E_{center} + E_{second}))$,

$A_1 = C_{[X,Y,G]} \times \sin(2\phi_1)$,

$A_2 = C_{[X,Y,G]} \times \sin(2\phi_2)$,

$C_{[X,Y,G]}$ are charge loss coefficients for grades 1–4 as above.

The charge loss coefficient depends only on the two pixels used in calculating A_i. Hence, even for charge loss correction to grades 5–8, we use the same charge loss coefficients as grades 1–4.

For all other grades (grade 0, grade > 12), we simply sum the $E+$ pixels to get $E_{corrected}$ without applying a charge loss correction. For example, for four pixel events (grades 9–12), the final energy is calculated as

$$E_{corrected} = E_{center} + E_{second} + E_{third} + E_{fourth}. \qquad (2.15)$$

The third and final step to calculate PI values is to correct for gain, and apply a fixed offset:

$$PI = E_{corrected} \times GAIN' + OFFSET_0. \qquad (2.16)$$

GAIN' values are tabulated for each grade for each pixel. PI are expressed as integer values in units of 40 eV. In addition, $OFFSET_0$ is subtracted from the "grade gain corrected energy" so that PI = 0 corresponds to an energy of 1.6 keV.

As mentioned before, the measured charge at pixel anodes is affected by the imaged hole signal. For photons with a given energy, as the depth of photon interaction in CdZnTe increases, the hole contribution increases, reducing the total energy. The common mode correction does not completely remove this effect. However, we can use the total hole signal in $E-$ pixels to estimate the depth of interaction. So, we calculate a scalar SURRPI to make it available for the science user for filtering on depth cuts:

$$SURRPI = SURR \times GAIN + OFFSET, \qquad (2.17)$$

where GAIN and OFFSET values for the central pixel are used.

After applying all these corrections, we obtain an events list with photon energies in PI units. Figure 2.9 shows laboratory spectra of radioactive elements measured by a NuSTAR CdZnTe hybrid. The uncertainty in energy arises primarily from three terms: electronic noise, Fano noise, and charge transport effects. Using the precision test pulsar in the ASIC, we have measured the electronic noise associated with the readout procedure to be ≈ 250 eV (FWHM) per pixel. This includes a contribution from leakage current flowing in `CdZnTe` due to the applied high voltage. Fano noise is the variation in $N_\pm$, the number of $e^- - h^+$ pairs created by a photon of fixed energy incident on the detector. The noise is typically less than noise expected from Poisson statistics. This is described by the "Fano factor" F, defined such that $\Delta E/E = \sqrt{F/N_\pm}$. For example, a 60 keV photon generates $N_\pm = 60$ keV/4.85 eV $\approx$ 12,400 $e^- - h^+$ pairs. Using $F \sim 0.1$ (Harrison et al., 2008; Niemela & Sipila, 1994), the uncertainty in the calculated photon energy is $\Delta E \approx 170$ eV. Charge transport effects arising from nonuniformity of the `CdZnTe` substrate dominate the uncertainty in energy reconstruction. Using only single pixel events ($\sim 1/2$ of all events), we attain an energy resolution of < 600 eV FWHM at 60 keV and 1 keV FWHM at 86 keV on NuSTAR hybrids. Including two-, three-, and four-pixel events in analysis slightly worsens the energy resolution.

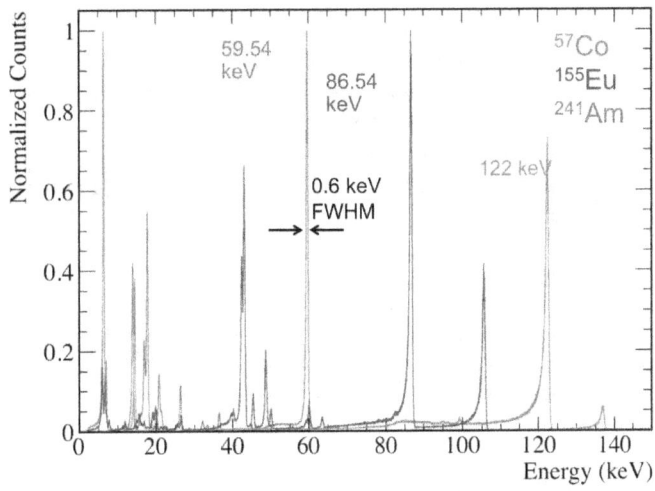

Figure 2.9. Radioactive source spectra measured by a NuSTAR `CdZnTe` hybrid. Independently acquired spectra for [57]Co, [155]Eu and [241]Am are plotted together, with peak counts normalized to unity. Only grade 0 events were used for making these spectra. All radioactive lines show a sharp core with some tailing towards lower energies. The core of the 59.54 keV [241]Am line has a FWHM of 600 eV. We obtain an energy resolutions of 400 eV at 6 keV.

The final step in event reconstruction is to calculate the detector coordinates for each photon events. This is done using the spatial probability distribution for event grades for each pixel, measured in ground calibration (Section 3.5). The probability distribution for event grades 0–12 is tabulated as a 7 × 7 grid in the Focal Plane Bench coordinate system, `DET1`. The `DET1` system uses integer coordinates with pixel size corresponding to $12''.3/5$ at the focal plane. For each event, we assign (`DET1X`, `DET1Y`) coordinates based on the probability distribution for the corresponding (`RAWX`, `RAWY`, `GRADE`) combination. These coordinates are later converted to the Optics Bench (`DET2`) using the mast aspect solution, and eventually to `SKY` coordinates.

2.6 Pileup

Pileup refers to two photons being incident on a detector in such a short duration, that they are read out as a single photon of higher energy. Pileup is not a major concern for pixelled detectors. However, for bright point sources, photons are incident at a high rate in a few pixel region of the detector, creating the possibility of pileup. Pileup was seen in laboratory measurements with the X-ray generator (Section 3.5) and the ^{55}Fe QE scans (Section 3.6).

Suppose a photon is incident on a detector pixel "A" at time $t = t_0$. We consider four time windows for the purpose of pileup:

1. *Trigger to lockout*: This time window spans the first $2 - 3$ μs from the asynchronous photon trigger to the MISC issuing a lockout signal to stop further triggers. A second photon incident in this duration on one of the 9 pixels to be read out will cause a hardware trigger. This is indistinguishable from a split pixel event arising from a photon with roughly the energy of both photons combined. This mode of pileup becomes important when the mean interval between photon arrival is a few times the coincidence window. In other words, $R_{\mathrm{pile}} \sim 1/(\mathrm{few} \times 2 \ \mu\mathrm{s}) \sim 10^5$ counts/s.

2. *Lockout to end of sampling*: In the interval from $(t_0 + 3 \ \mu\mathrm{s})$ to $(t_0 + 8 \ \mu\mathrm{s})$, the hybrid will not issue any hardware triggers. However, if another photon is incident on one of the 9 pixels which will eventually be read out, then charge deposited by the photon still contributes to the POST sample sum for that pixel. Such events start occurring at count rates a few times lower than R_{pile}. They can be screened out in careful postprocessing by searching for pixels which have enough energy deposited to cause a hardware trigger, but did not actually have a hardware trigger. Note that a few pathological cases are possible where this screening procedure will not identify such events. For example, if the second photon was incident next to the trigger pixel at $t \approx t_0 + 6 \ \mu$s, it may raise the POST sum to a level between the hardware and software trigger thresholds. This will be missed by the screening procedure but will raise the measured energy of the event (pileup).

3. *End of sampling to just before detector live*: Any photons incident from $(t_0 + 8 \ \mu\mathrm{s})$ to $(t_0 + 2.5 \ \mathrm{ms} - 8 \ \mu\mathrm{s})$ neither trigger the hybrid, nor contribute to the energy deposited in the event.

4. *Last 8 μs of dead-time*: Consider the case where photon "P" is incident on pixel "A" within 8 μs of the detector going live. The PRE sum for that event contains some capacitor samples while the detector was still dead. If a photon "Q" was incident on one of the 9 pixels (say, pixel "B") during those samples, then it would not trigger the detector. But, the charge deposited by that photon still contributes to the PRE samples for that pixel. Such events start occurring at count rates a few times lower than R_{pile}, and their consequences are slightly complicated. If photon P does not deposit any energy in pixel B, then the POST $-$ PRE value for this pixel will actually be negative. This will *add* some extra energy to PHA of the triggered pixel A when we apply the common mode correction (Equation (2.11)). Like case 2 above, this can be screened out in postprocessing. On the other hand, if photon P deposited significant charge in pixel B, then the measured POST $-$ PRE value will be lower than the true energy deposited in that pixel. Whether this can be screened out or not depends on the energy deposited by both photons in pixel B.

Most astrophysical sources will have count rates significantly lower than R_{pile} and these effects can be safely ignored. However, we will also carry out simulations to estimate the magnitude of these effects for NuSTAR observations of bright point sources.

Attaining the best performance from each hybrid requires careful calibration in the laboratory, a procedure which took a few hundred hours per flight hybrid. In the next chapter, I discuss the details of this calibration procedure.

Chapter 3

Calibration

Let us follow the optical path of an astrophysical source photon as it makes its way to a NuSTAR detector. The photon first passes through the optics thermal cover. Then it is reflected twice by the optics, and passes through the back thermal covers. The optics reflectivity has a sharp cutoff at 79 keV due to the K-edge of Pt. Closer to the focal plane, there are baffles which stop most of the stray light from near the field, which did not make it through the optics (Figure 3.1). The photon then passes through Be windows mounted on top of the detectors. These windows absorb lower energy photons ($\lesssim 3$ keV)—this is important to keep the count rate reasonable by letting only photons in the energy range of interest reach the detectors. Finally, the photon interacts with the CdZnTe crystal and triggers a pixel. This initiates readout, and a photon event data packet is created and handed off to the spacecraft.

Each element of NuSTAR that a photon interacts with needs to be calibrated with great care in the laboratory to fully characterize the performance of the telescope. The degree of certainty in calibrations reflects directly on the quality of scientific output of the mission. In Section 3.1, I describe the performance requirements for NuSTAR and how they lead to calibration requirements at the component level. In Section 3.2, I discuss the calibration steps for every hybrid. In the remainder of this Chapter, I discuss all steps that I executed as a part of the hybrid calibrations, followed by transparency calibrations for the Be windows and optics thermal covers.

3.1 Requirements

Calibration requirements for NuSTAR are laid out from the primary science objectives. In brief, the relevant requirements are:

- **Absolute flux calibration**: For a source with a power-law spectrum with a photon spectral index of 1.7, the systematic uncertainties in flux should be less than 30% (3-σ) over the central $11' \times 11'$ of the field of view (FOV), and less than 15% in the 6–10 keV band in the central $2' \times 2'$. In order to accurately measure ^{44}Ti lines, the systematic uncertainty in absolute flux measurements should be $< 15\%$ in the 60–79 keV band in the central $8' \times 8'$.

- **Spectral calibration**: For a source with a power law spectrum of the form $F = kE^{-\alpha}$, $-1 < \alpha < 1$ the systematic error in measuring α should be less than 0.15 (3-σ) in the 6–10, 10–30, and 30–79 keV bands over the central $11' \times 11'$ of the FOV. In the 60–79 keV band, the uncertainty should be < 0.15 in the central $8' \times 8'$ of the FOV.

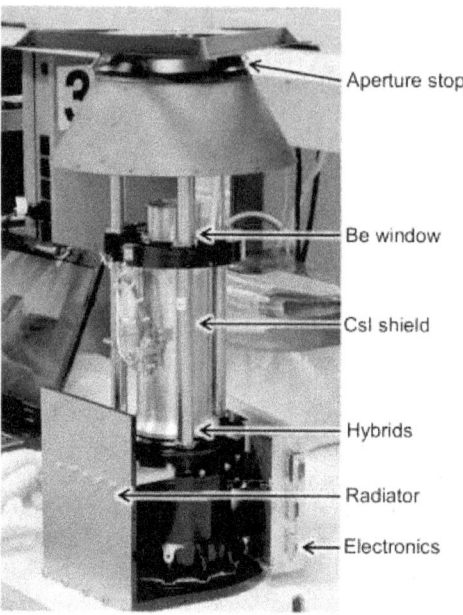

Figure 3.1. A NuSTAR focal plane module. The deployable aperture stop (top, black) sits on three telescoping rods. A copper can holds the Be windows, which block low energy photons. A CsI anti-coincidence shield surrounds the assembly, and a photomultiplier tube under the hybrids monitors any scintillation events in the shield. The hybrids are enclosed near the base of the golden assembly. On the left, a radiator passively cold-biases the entire module, heaters near the hybrids maintain operating temperature ($\sim 5°$C). Control electronics are mounted on the right.

- **Point spread function characterization**: The integrated value of the instrument Point Spread Function (PSF) from 70%–90% encircled energy should be determined to 10% (3-σ) over the central $11' \times 11'$ of the FOV. The radial and azimuthal variations of the PSF (70% encircled energy) should be characterized to 3% and 10% (3-σ) accuracy respectively, over the central $11' \times 11'$ of the FOV.

The instrument level requirements derived from these calibration requirements are listed in Table 3.1. This is the net performance requirement for NuSTAR, including ground-based and in-orbit calibrations, in order to meet the science objectives. The motivation for defining the energy range for NuSTAR is to obtain hard X-ray spectra that complement existing soft X-ray data, without leaving behind any gap in the spectrum. This desire to have a sufficient energy overlap with soft X-ray missions drives our calibration requirements at the low energy end (6–10 keV). At the highest energies (60–80 keV), one of the science goals is to study the ^{44}Ti line emission in young supernovae remnants. To properly analyze the lines, we need to characterize the underlying continuum well, and in turn must understand the telescope response with high accuracy with good energy resolution. Requirements on characterizing the PSF stem from the aim of accurately understanding point images to study diffuse emission around them in the galactic plane.

It is impossible to measure the performance of the instrument at every possible combination of energy, incident angle, etc. Instead, we have very detailed models and simulations for each component of NuSTAR. The aim for ground and in-orbit calibration is to obtain select data sets to fit for a few model parameters, and verify the model. For example, the optics are illuminated with an intense X-ray beam at a range of

Table 3.1.   Instrument-level calibration requirements

Topic	Requirement	Rationale
Alignment		
Optical Axis Knowledge	$15''$	Throughput determination
Effective Area		
Absolute effective Area: 6–10 keV (central $2' \times 2'$)	15%	Cross calibration with low-energy missions
Absolute effective area: 6–10, 10–30, 30–79 keV bands ($11' \times 11'$)	25%	Hardness ratio determination and surveys
Absolute effective area in each 2 keV bin between 60 and 80 keV	12%	^{44}Ti yield measurement
Relative effective area in each 2 keV bin between 6 and 79 keV over central $8' \times 8'$	5%	Spectral index fitting and bright sources
Relative effective area in each 2 keV bin in the 60–79 keV range in central $8' \times 8'$	3%	Continuum modeling and subtraction for ^{44}Ti
Point Spread Function		
Integrated PSF 70–90% encircled energy over $11' \times 11'$	10%	Mapping diffuse features/point sources
PSF as function of radius out to 70% encircled energy	3%	Flux determination. Remove point sources in diffuse emission
PSF as function of azimuth out to 70% encircled energy over $11' \times 11'$	10%	Mapping diffuse extended features and jets

angles, and the data are used to fit and verify a raytrace model. For the hybrids, the event reconstruction described in Section 2.5 are based on working knowledge of the electronics, and a charge transport model for CdZnTe.

The high level calibration requirements translate into the following requirements that apply directly to focal plane calibration.

1. The uncertainty in position bias correction in the measurement of the X-ray interaction relative to a physical detector coordinate system shall be less than 100 μm anywhere on the active area. This pertains to the systematic part of the positioning due to detector distortions.

2. The Quantum Efficiency of each focal plane hybrid detector shall be determined to 5% accuracy in the 6–80 keV range.

3. The photopeak efficiency shall be determined to 3% accuracy in the 6–80 keV range.

4. The transparency of the Be entrance window shall be determined to 0.5% accuracy between 6 and 80 keV.

5. The focal plane/electronics system shall measure the absolute energy of an X-ray to better than 0.5 keV (3-σ) from 10 to 60 keV.

Apart from these five requirements for the Focal Plane Module (FPM), we also worked to verify the transparency of the optics thermal covers at low energies. The transparency data fold into the overall uncertainty of the optics throughput at low energies.

3.2   Detector Screening, Selection, and Calibration Steps

NuSTAR flight hybrids were made by the Space Radiation Laboratory (SRL) at Caltech, spanning the full range from design to testing and calibration. The life of hybrids begins with selecting the right ASIC and crystals. The manufacturer of CdZnTe (EI Detection & Imaging Systems, Saxonburg, Pennsylvania) provided us with IR diffraction images of the wafers, to select suitable uniform regions to be diced into NuSTAR CdZnTe crystals. These crystals were then taken to Brookhaven National Laboratory (BNL) for X-ray diffraction imaging. The diffraction images were examined to identify any crystal defects which could degrade charge transport properties of the detector. The best crystals which passed these inspections were selected for attaching anode and cathode contacts. A final selection step was to measure the bulk and surface conductivities as a proxy for leakage current. High leakage crystals were rejected. In parallel, ASICs from ON semiconductors were tested for basic functioning. They were then subjected to "burn in"—an accelerated device aging test by heating at high temperature for one week. The ASICs were tested again and ones with low number of disfunctional pixels and no degradation through bakeout were selected as flight ASICs. The best CdZnTe crystals and the best ASICs were bonded to form candidate flight hybrids. At this stage, we commenced the selection and calibration of eight hybrids to be flown on board NuSTAR (Figure 3.2). Some selection steps were interspersed with calibration, as hybrids showing poor performance were demoted from "flight" to "backup" status.

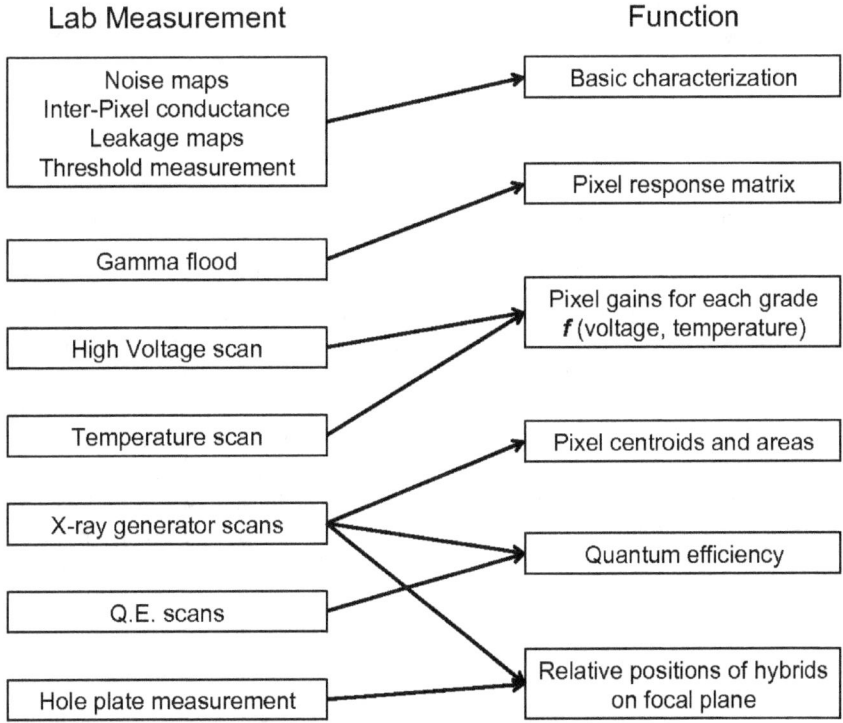

Figure 3.2. Overview of the calibration procedure. The left column lists various measurements undertaken in the laboratory. The right column lists the requirements and "measureables" for calibration. Calibration steps may be related to multiple requirements, and vice versa.

3.2.1 Hybrid Selection and Screening

The first screening step is to shine a radioactive source like ^{241}Am on a hybrid and generate an output spectrum. This allows to verify the functioning of the hybrid including noise performance, number of dead pixels, and approximate energy resolution. Next, we undertake more detailed characterization of the noise to make noise maps. We also measure the interpixel conductance (IPC) and reject any hybrids with high IPC values. To improve energy resolution, it is desirable to operate hybrids with higher values of potential difference between the anode and cathode. However, increasing the high voltage (HV) increases the leakage current in the pixel, and thus the baseline charge deposited in each sampling capacitor in a sampling interval. This, in turn, increases the probability of getting false triggers due to a voltage above the trigger threshold. We measure the leakage current for each hybrid to select its optimal operating HV. The measurement is repeated at various temperatures, to verify stability around the operating point. Based on these characteristics, we identify the few noisiest pixels in each hybrid, and disable them during regular operation.

Pixels are triggered when the charge deposited exceeds a certain threshold value. This threshold value is set by an external Digital-to-Analog Converter (DAC) on the motherboard. We want to set the threshold value as low as possible, to detect lower energy photons. However, this also increases the spurious noise triggers, decreasing the instrument live-time. Hence, we test various threshold values to determine the lowest possible setting where the noise contribution remains negligible, with less than ten noisy pixels disabled.

3.2.2   Pixel Response Calibration

As discussed in Section 2.5, we need to establish a channel to keV conversion for each pixel and grade combination. We calculate this mapping using X-ray line emission from the radioactive isotopes ^{57}Co, ^{241}Am, and ^{155}Eu (Figure 2.9). To obtain sufficient statistics to calculate these conversions, we undertake "γ flood" integrations for 48 hours for ^{241}Am and 24 hours each for the other two isotopes.

Several aspects of the functioning of the hybrids depend on extrinsic factors. The electronic noise and pixel gains vary with operating temperature. At higher temperatures, the leakage current in CdZnTe increases and can effectively saturate the readout. The ASICs do not function properly at very low temperatures. We take ^{241}Am spectra at varying temperatures to measure these variations and determine the operating temperature. It was seen that the hybrids perform best at $\sim 5°$ C, well within the temperature range attainable in-orbit. Leakage current and pixel gains depend on the applied HV, too. Hence, we take ^{241}Am spectra at the operating temperature with different HV settings. It was found that the optimal HV is -400 V for four of the flight hybrids and -450 V for the other four. Data from these HV scans are also analyzed for measuring the $\mu\tau$ products for electrons and holes in CdZnTe, to be used in modeling the detector charge transport properties.

3.2.3   X-ray Pencil Beam Scan

Pixels in NuSTAR hybrids are formed the spacing anode pads and are not physically insulated from each other (Figure 2.4). From X-ray diffraction images, we know that the CdZnTe crystals have various defects which will deform the electric fields. Apart from these impurities, we also have to map out which regions of the hybrids preferentially give single or multiple pixel events. With this aim in mind, we scanned every hybrid with a fine X-ray pencil beam. From charge transport simulations, we expect that split pixel events will typically occur roughly in a hundred micron wide region near pixel boundaries. To effectively map this, we selected a beam size one-tenth the size of the 605 μm pixel. These scans are the first ever to probe the structure and functioning of CdZnTe detectors at such fine scales. The design, setup, and execution of these scans forms a large part of my thesis work, and will be discussed in detail in Sections 3.3 — 3.5.

3.2.4   Quantum Efficiency Measurement

The final step in calibration of the hybrids is measurement of the absolute quantum efficiency. The modus operandi for QE measurements is to take a well-calibrated radioactive source, and measure its total fluence using a NuSTAR flight hybrid. The ratio of the counts detected in the hybrid to the known source intensity is the QE. In practice, we measured the QE independently at various energies and for different regions of each hybrid. These measurements of average QE of regions of the hybrid are augmented by the X-ray scan data to calculate the absolute QE of each pixel. QE measurements are discussed in detail in Section 3.6.

3.3   The X-ray Generator Laboratory

I designed and implemented the X-ray Generator (XRG) Laboratory setup for undertaking the X-ray pencil beam scans and QE measurements. Later, we also adapted the setup for transparency measurements of the flight `Be` windows and optics thermal covers. Let us look at the essential details of the XRG lab to obtain context for the sections to follow.

3.3.1 Hardware Design

The primary hardware component of X-ray pencil beam scans is an X-ray generator with a collimator to produce the fine beam. For scanning, it is practical to translate the hybrids in front of a stationary X-ray beam. The enclosure for the hybrids should maintain operating conditions. External hardware provides power, data, and control signals. The same setup should allow replacement of the X-ray beam with a radioactive source for QE measurements.

The workhorse instrument is a 3 kW X-ray generator by Rigaku Corporation.[1] We use the XRG with a Molybdenum (`Mo`) tube, obtaining a Bremsstrahlung spectrum superposed with `Mo` K lines. The X-ray tube is housed in the XRG tower—an `Al` cylinder lined with lead on the outside to contain the intense radiation. A 12×4 mm aperture produces a divergent beam (opening angle $\sim 45°$). We collimate this beam down by using a pinhole at the end of a 43 cm beam tube (Figures 3.3, 3.4). Due to space restrictions, the beam tube has to be placed a short distance from the tower aperture. To contain any radiation that may leak from this space, we added an acrylic cylinder with lead sheets on the inside to contain all scattered radiation. A second layer of safety is provided by leaded glass panels with interlocks, which switch off the X-ray beam if opened. Lastly, the side of the enclosure which directly intercepts the beam has thick `Al` plates. The total radiation leak from the entire setup under operation is undetectable for our "low energy" configuration, and under 0.07 mR/hr for the "high energy" configuration.

The design goal is to obtain a beam of ~ 60 μm at the detectors. Space constraints and safety considerations dictate that the detector surface is at least a few centimeters from the pinhole. With the small desired spot size and relatively short beam length, beam divergence becomes a serious consideration. One way to control the beam size is to add multiple collimating pinholes along the beam path. However, this significantly decreases the source flux and makes alignment extremely difficult. Instead, we choose to account for the divergence in design by using a smaller pinhole and using the minimum practical pinhole–detector distance. This leads to a desired pinhole diameter of ~ 40 μm, surrounded by material thick enough to block the intense X-ray beam. The pinhole needs to be elongated in one direction to compensate for the rectangular aperture of the X-ray tower. We accomplished this by using a pair of crossed slits, made from tinned lead. The slit widths were tweaked to the desired size under a microscope and the blades were then clamped down by screws. After multiple iterations, I achieved a slit size of 30×50 μm. Using a raytrace model, I calculated that the spot size on the detectors is 70×50 μm (FWHM, Figure 3.5).

We modify the same setup for QE scans. A slot cut in the X-ray beam tube can be opened to insert a radioactive source holder. The three calibration sources (^{57}Co, ^{241}Am, ^{155}Eu) were mounted in their respective holders throughout the calibration process, to minimize variations of experimental conditions. A pin on the source holder ensures accurate source positioning to better than a mil (25 μm). The fine slit is removed and replaced with a "QE mask": a sheet of tinned lead with a precisely milled square hole at the center. In this configuration, the radioactive source illuminates a 12×12 pixel grid on the hybrid.

The hybrids are mounted on a motherboard screwed in vertically inside a "cold box" (Figure 3.4). X-rays are incident on the hybrid through a Mylar film. The box is cooled by pumping in cold, dry air from a forced air unit through an opening on the side. The box has 1/2" to 1" thick insulating foam on the inside to decouple it thermally from the surroundings. Air flowing directly onto the hybrid can excite microphonic vibrations in the detector, which result in increased electronic noise. To counter this

---

[1]http://www.rigaku.com

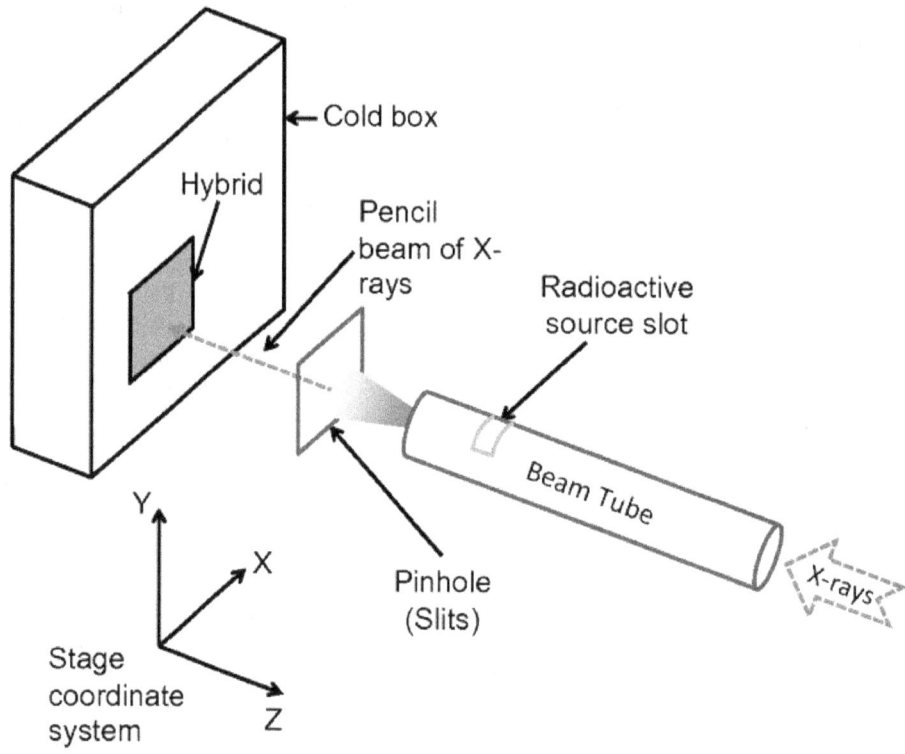

Figure 3.3. Schematic of XRG setup. A hybrid (gray) is mounted vertically in a cold box that can be translated on X-Y stages (not shown). The stage coordinate system is shown in the lower left. X-rays are generated by an X-ray tower on the right (not shown). The divergent X-ray beam travels along the $-Z$ direction through horizontal beam tube (blue) to a pair of slits forming a pinhole. The pinhole forms a mildly divergent pencil beam which is incident on the hybrid. For "QE scans," a radioactive source is inserted in the beam pipe, and the pinhole is replaced by a square mask. A photograph of the setup is shown in Figure 3.4.

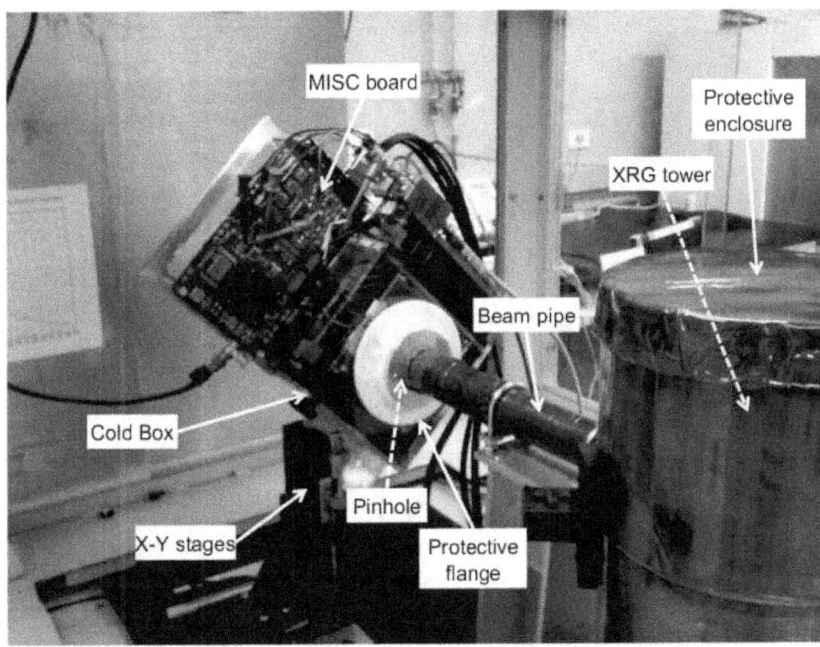

Figure 3.4. Photograph of XRG setup. X-rays are generated by an X-ray tower contained within a protective cylinder lined with lead (gray, right). X-rays travel through horizontal beam pipe to a pinhole concealed beneath the protective flange (white). Hybrids are mounted vertically in the controlled environment of the "cold box." The cold box is mounted on X-Y stages for translation. A MISC board with external connections is seen on the upper left of the cold box.

problem, we have a diffuser which sends the cold air in through several small apertures such that no stream is directly incident on the hybrid. The exit hole is diagonally opposite the entrance hole to ensure proper air circulation inside the cold box.

We use a pair of Newport LTA-HS translation stages to scan the cold box and the hybrids across the X-ray beam. These stages have an absolute accuracy of ± 7.5 μm and are positional repeatability of ± 1 μm. The 50 mm stage traverse leaves a few millimeter margin over the desired range for scanning all four hybrids mounted together on the motherboard. In practice, scheduling issues led to calibration of one detector at a time. The stages are controlled by a Newport ESP-300 stage controller, commanded by custom software on a PC (Section 3.3.2). The entire assembly is designed so that the mounting of the detector relative to the beam (including joints to motherboard, cold box, stage, optics plate) is repeatable to about a millimeter. The final position is tweaked by checking the beam position in software, and issuing an offset command to get the beam to an expected location. Hardware and software limits ensure that there is neither excess strain in connectors, nor any physical risk to flight hardware due to erroneous stage operation.

For reasons discussed later, the lab humidity is usually maintained at around 50%. When hybrids are cooled for testing, condensation develops on the outside of the cold box. We employed several measures to mitigate any risk of water seeping to the hybrid. First, the cold box is designed and mounted so that any liquid drip path completely bypasses the hybrids. Second, and more importantly, we decided to put

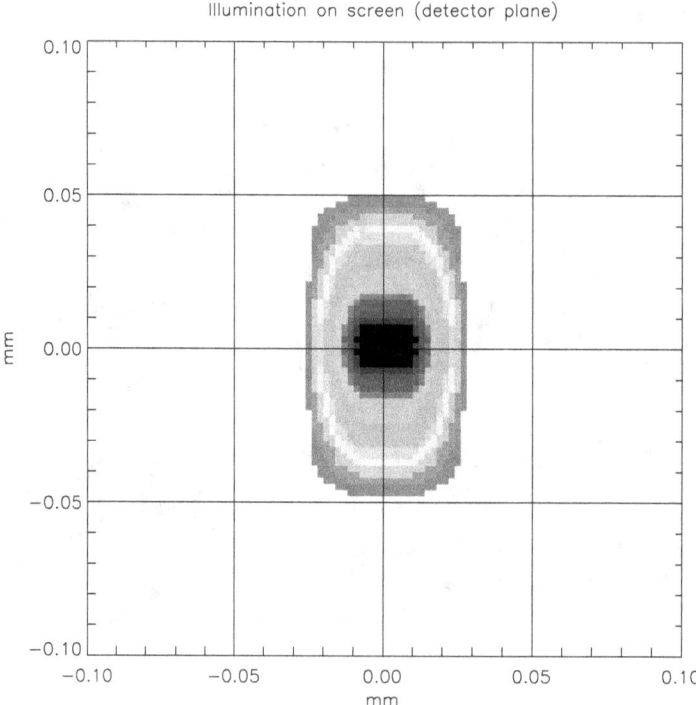

Figure 3.5. Simulated profile of the fine XRG beam on a detector. The beam is collimated by a $30 \times 50~\mu$m pinhole.

the entire apparatus in a nitrogen environment in a purge box (Figure 3.6). Before cooling down the apparatus, we start the N_2 flow to purge out the vapour–bearing room air from the box. The purge box is designed so that opening and closing it is quick: this allows us to make minor modifications inside the box (like swapping radioactive sources) in a short duration and close the box again without any risk of getting condensation on the cold box.

A major concern in handling our hardware is electrostatic discharge (ESD). The potential difference in such discharges can be as high as kilovolts, and can potentially damage the electronic components. Even partial damage to any electronic components is a high risk, as the component may fail during the mission lifetime in orbit. We take several steps to mitigate the risk of ESD in the lab. All equipment are connected to a common ground to ensure that they are at the same potential. Ambient humidity is maintained at near 50% while handling all electronics, as dry air greatly increases the risk of ESD. Lastly, we take steps to avoid discharge spikes. Insulating surfaces are conducive for large buildup of charge, while conductive surfaces facilitate rapid discharge. To avoid both these cases, most lab surfaces are specially coated to be static dissipative: such surfaces dissipate charge at a low enough rate to not risk any hardware damage. All lab operators are trained in ESD safety precautions as per JPL guidelines.

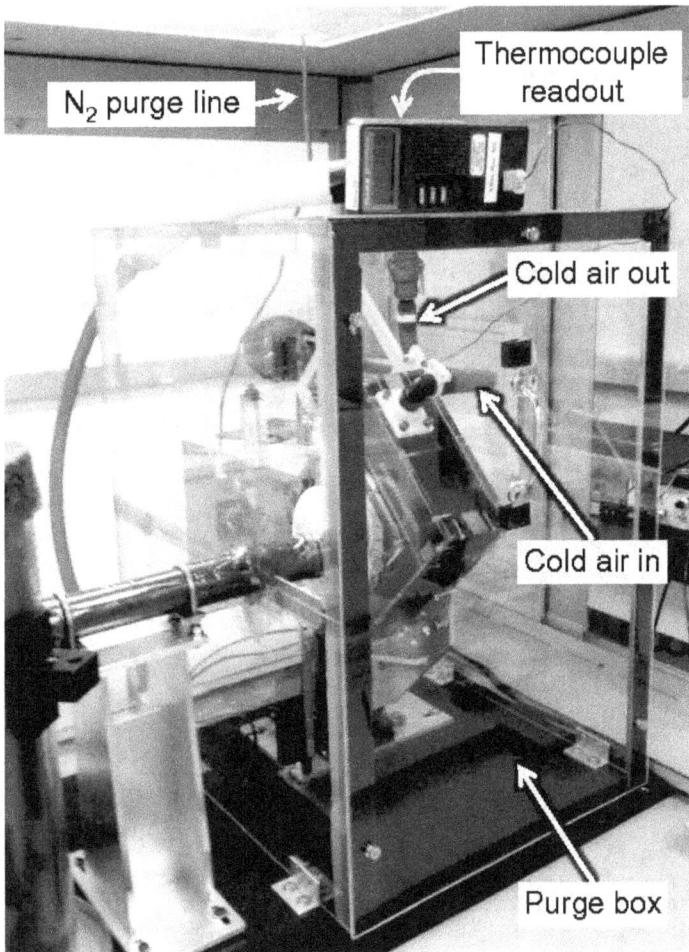

Figure 3.6. Cooling and purge systems for the XRG setup. Cold, dry air from a forced air unit (not seen) enters the cold box from the right and flows onto the hybrid through baffles for diffusing the air flow. Another air pipe leads the air out through the top of the enclosure. The transparent enclosure is the purge box. N_2 boiloff from a liquid nitrogen tank is fed into the box by a copper pipe (top). The gas diffuses out through spaces in the box, maintaining a dry environment inside.

3.3.2 Control Software

The X-ray generator setup is controller through custom software. The basic requirements of the software were

1. Command interfaces to the ESP300 stage controller and the NuSTAR ASIC,

2. Automation of XRG and QE scans,

3. Combine stage and ASIC data into a flight-like format which can be processed by existing analysis

software,

4. Ability to remotely control and monitor operation,

5. Complete and comprehensive logging of all hardware and user interaction, even in case of crashes,

6. Ability to manually override any steps if required,

7. Flexibility to execute test cases.

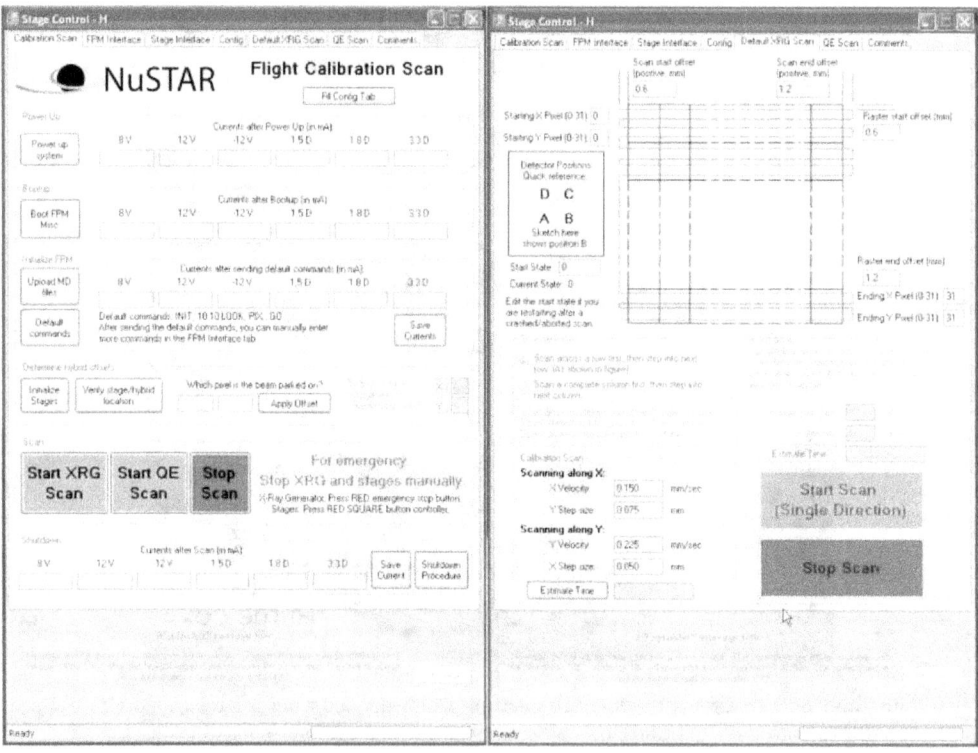

Figure 3.7. StageGUI control software. The principle "Calibration Scan" tab (left screenshot) guides a user through powering up the hardware and establishing stable operating conditions. Individual tabs allow direct command access to the ESP300 stage controller or to the hybrid. Other tabs are used to set up the XRG scan (right screenshot) or QE scans. The config and comments tabs are used for logging information.

To take advantage of existing code and availability of drivers, we decided to use Microsoft Windows to interface with the hardware. To execute the calibration scans, I developed a graphical user interface (GUI) in VC++ (Figure 3.7). This custom StageGUI software interfaces with the hybrid as well as the stage controller. The hybrid interface utilizes two RS422 ports. First, the MISC is initialized and booted over the command port. The bidirectional port is then used for sending commands in the forth language and monitoring responses. The other port is the unidirectional data port, which is continuously monitored for event data. For test and debugging purposes, a WinForth terminal on the PC can be used for interfacing with the hardware. The ESP300 stage controller provides a fully functional command interface for controlling any connected translation stages. We use the RS232 serial interface to the controller. To ensure repeatability of scans, stageGUI initializes all basic parameters of the stages, including units, software limits, velocities, and accelerations.

A typical XRG or QE scan begins with a default bootup procedure, followed by interactively commanding the hybrid to establish stable working conditions. The operator then sets the parameters of the scan to define the movement pattern for stages. StageGUI controls the movement of the hybrid by commanding ESP300 appropriately. Event data from the hybrid is recorded as a continuous stream. At predefined intervals, StageGUI queries the stage position and adds a stage position packet to the data stream. The stage packets follow the NuSTAR data format, so that any existing analysis software simply ignore these packets and are still able to process data.

All user interactions and automated commands are logged to the screen and written directly to disc without buffering. This ensures traceability of crashes and issues. Debug modes can be invoked to override the default command sequence. Toggling of debug modes is logged as well. Remote monitoring is achieved by using the Teamviewer software, with redundant access is obtained by using the logmein service.[2] Data are periodically transferred to sarasvarti, the main NuSTAR server.

3.4 Calibrating the Setup

We undertook extensive characterization of the setup to qualify it for calibrating NuSTAR flight hardware. For XRG scans, the primary requirements are measurement of the pencil beam profile and verifying that the XRG flux is constant over the duration of a scan. For the QE setup, we had to measure the fluence of each source with high accuracy. Here I describe how I accomplished each of these goals. We also tested the reproducibility of various quantities calculated from scans. These tests are discussed later, with the analysis of respective scans.

3.4.1 Beam Shape

As seen in the raytrace (Figure 3.5), we expect the pencil beam to be symmetric in both X and Y directions. We measured the X and Y profiles of the beam with a knife edge scan. We place a detector at some distance from the slit such that it intercepts all the X-rays coming from the slit, then move a knife edge to progressively occult greater fractions of the beam, and measure after each step of the knife edge. The difference in count rates between successive steps is the count rate in the obscured part of the beam. We used an Amptek Si detector with a circular input collimator of 2.38 mm diameter. We know from raytrace modeling that this detector will intercept the full X-ray beam even at several inches from the slits. Since we want to measure the beam profile at the fiducial CdZnTe surface, the knife edge is kept in that plane, while the Si detector is mounted further away from the slit. We acquired data by stepping the knife edge by small amounts ($\sim$10 μm) between integrations. We measured the beam FWHM to be $\approx$50 μm in X and $\approx$70μm in the Y direction (Figure 3.8). Consistent results were obtained on repeating the knife edge scan after several months, after calibration of flight hybrids.

---

[2]http://www.logmein.com

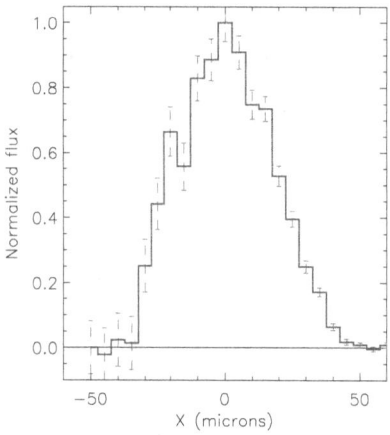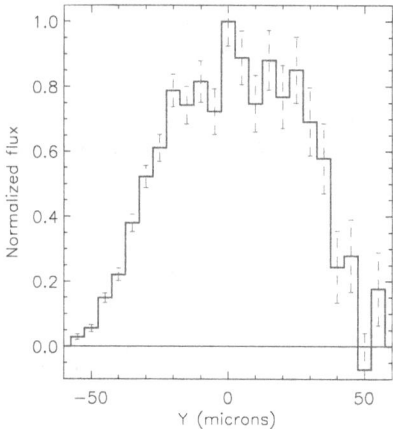

Figure 3.8. XRG beam profile with a knife edge scan in the X (left panel) and Y directions (right panel). The blue
          lines show the differential count rate measured by moving the knife edge in steps of 5 μm, normalized
          such that the peak response is unity. The dashed error bars show Poisson uncertainties in fluxes. The
          physical coordinates are set to be 0 at the peak intensity of the beam. The measured beam FWHM (Δ)
          from multiple knife edge scans is ΔX $\approx$50 μmand ΔY $\approx$70 μm.

3.4.2 Rate Stability

The rate stability of the XRG was tested in two ways: first, with a Si detector and a scalar, and then with
NuSTAR detectors.

I set up the X-ray generator (XRG) with a 50 × 70 μm slit, which generates a spot size of 100 × 100 μm
at the detector. I used a Mo tube, operated at 45 kV, 20 mA. Data were acquired using the Amptex Si
detector and a scalar. The background rate for this detector is negligible: ∼0.03 counts/s. I aligned the
Si detector to the XRG beam using the MCA for quick readout. The detector position was adjusted to
achieve maximum count rate. Then I swapped the MCA with the scalar. The measurements were done
by manually starting and stopping the scalar acquisition as per a stopwatch. The time recorded on the
stopwatch (≃10 s) was used to calculate count rates. The RMS scatter in the time intervals is 0.1 s. I took
50 readings for the XRG, at a count rate of about 400 counts per second. For comparison, I also obtained
20 readings with an ^{241}Am source, with distance adjusted to get a similar count rate.

The measured counts for the ^{241}Am have a mean of 440 counts/s. The scatter in readings is 6.6 counts/s—
consistent with a Poisson distribution. This test shows that the Si detector and scalar combination gives a
reliable measurement of the count rate. For the XRG, the mean count rate is 412 counts/s, with a standard
deviation of 6.4 counts/s. This is consistent with the expected scatter for a Poisson distribution.

Next, I conducted a stability measurement using the NuSTAR detectors themselves. I set up the XRG
at 20 kV, 2 mA, and placed H78 in a the cold box to intercept the beam. The stages were manually
adjusted so that the beam was placed close to the center of a pixel. I measured a count rate of about
333 counts/s. Data were acquired for ∼22 h, yielding about 79,000 one-second count rate measurements.
It was observed that the count measured by the detector was constant over the entire run. Due to the
dead-time interval associated with reading out each event, the count-rate distribution is a modified Poisson
process (Figure 3.9). The observed data are in excellent agreement with a simulated count rate distribution

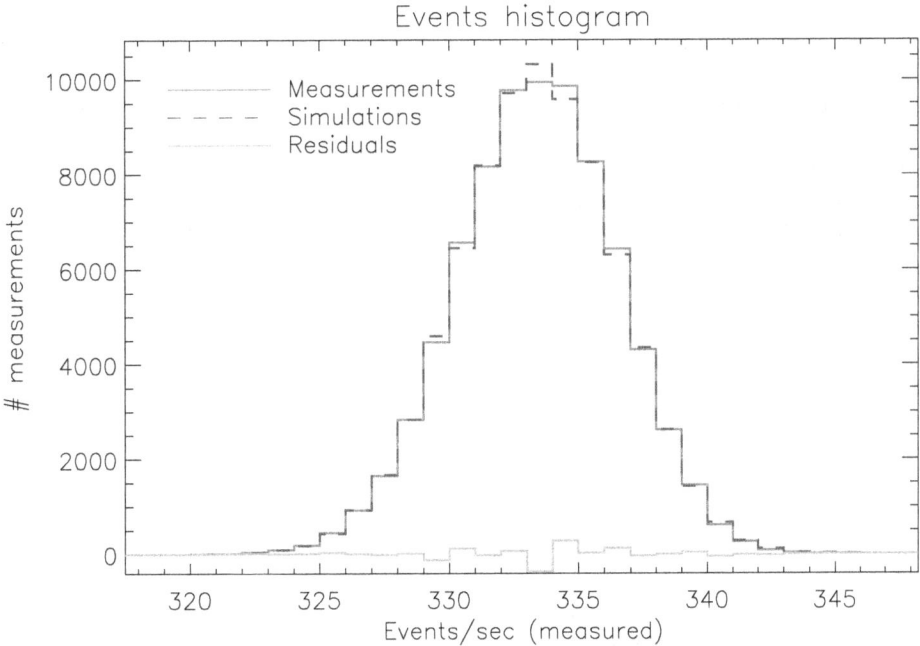

Figure 3.9. XRG stability measurements. The red curve is a histogram of counts per second obtained in a 22 hour integration. The mean of the histogram is sensitive to the input count rate, while the width depends on the dead-time per event. The dashed blue histogram shows a simulated distribution adjusted for the input count rate, using 2.5 ms dead-time per event. The residuals (green histogram) demonstrate the good quality of the fit.

for an incident beam with constant flux. The mean of this distribution is sensitive to the total count rate, while the width depends on the dead-time per event. Using this data set, I verified that the dead-time per event is 2.50 ms.

3.4.3 Radioactive Source Fluence

As explained in Section 3.2.4, the quantum efficiency measurements consist of shining a radioactive source with a known count rate on a part of the detector, and calculating the QE from the measured count rate. A prerequisite for this is knowing the source flux well. We accomplished this by measuring the source count rate with good statistics, using a Ge detector. We have a detailed model for the response of the Ge detector from previous work. We placed each radioactive source in turn in the beam tube with the QE mask, and placed it such that the distance from the Ge detector was same as the distance from the CZT detector during calibrations. The calibration Mylar window covering the cold box was also added in the beam path. In this configuration, the Ge detector intercepts the entire beam coming from the QE mask. During calibration, the source and detector were both inside the purge box, in a nitrogen environment. In the ≈2.5 in gap between the source and detector, air absorbs about 15% radiation at 6 keV, while N_2 absorbs only about 12%: a significant difference. Ambient humidity further increases the absorption in air.

So, we placed the entire setup in a nitrogen bag. We then took long integrations to beat down Poission noise, and measured source fluxes with high precision. The count rate of the ^{55}Fe ($\sim$40,000 counts/s) is too high for the Ge readout electronics. We could not revert to the Amptex Si detector since its active area there is too small to intercept the full beam. Hence, we decided to insert an attenuator to reduce the count rate on Ge to about 180 counts/s. In turn, we would calibrate the attenuator itself using the Si detector, by simply taking an I/I_0 pair of measurements with and without the attenuator respectively. With a combination of these steps, we calculated the final fluence of the 3 calibration sources through the QE mask to be $\approx$150 counts/sfor ^{155}Eu, $\approx$60 counts/sfor ^{241}Am, and $\approx$30,000 counts/sfor ^{55}Fe. The uncertainties in these measurements were propagated through to the uncertainty in-flight detector quantum efficiency measurements.

3.5   Pixel Centroids and Areas

We use X-ray generator scans to measure the spatial response of hybrids to X-rays and calculate the centroid and area of each pixel. I describe the scan configuration and procedures in Section 3.5.1, followed by data analysis and results in Section 3.5.2.

3.5.1   Procedure

In the XRG configuration, we mount the fine slit (30 $\times$ 50 μm) in front of the beamline. A hybrid is mounted in the cold box on a flow bench. We mount the cold box on the translation stages and commence the cooldown procedure. The operators follow a hardcopy checklist throughout this process. When the hybrid stabilizes at operation conditions, we check the position of the X-ray beam and recenter it as needed. Throughout our calibrations, we found that the absolute position of hybrids was repeatable to about a millimeter.

We undertake two types of scans: a Low Voltage (LV) scan and a High Voltage (HV) scan (Table 3.2). For the LV scan, we operate the XRG at 20 kV, generating a relatively "soft" spectrum. The e-folding path of a 15 keV photon in CdZnTe is merely 36 μm, so most of the photon interactions occur very close to the cathode. The charge cloud moves through the full depth of the CdZnTe crystal, so LV scan results reflect charge transport properties of the entire detector. Higher energy photons penetrate deeper into CdZnTe. The electron cloud generated by such an interaction will be unaffected by any crystal defects close to the cathode. To study this effect, we repeat the scans in HV mode with the XRG set to 60 kV (Figure 3.10). To avoid being dominated by the LV counts, we add a 2.3 mm Al slab in the beam path which absorbs the low energy component.

A typical scan consists of a full raster across the detector, once across rows and once across columns. The beam is moved across the full detector with some margin so that it definitely crosses over the edge of the detector. We call this the scan direction. Then we move it sideways (raster direction) by a step equal to the beam size in that direction, and then move it back over the detector to the other side (Figure 3.11). Such a scan is very effective in determining pixel boundaries perpendicular to the scan direction. In order to complete the mapping of pixels, we repeat the scan with raster and scan directions switched. Based on exact scan settings, we get ten to fifteen thousand counts in each pixel.

The NuSTAR requirement *L4-FPE-65* states, "The uncertainty in the position bias correction in the measurement of the X-ray interaction relative to a physical detector coordinate system shall be less than 100 microns anywhere on the active area." This can be broken down into two sub-requirements: first, determining the centroids of each pixel on a hybrid, and second, measuring the relative positions of all

Table 3.2.   X-ray generator settings for hybrid scans

Mode	Voltage	Current	Incident Count Rate	Measured Count Rate
	kV	mA	counts s^{-1}	counts s^{-1}
LV	20	2	$\approx$2000	$\approx$330
HV[a]	60	2	$\approx$10000[b]	$\approx$385

[a]In HV mode, a 2.3 mm thick Al plate was added as a "filter" to decrease the total count rate by preferentially blocking lower energy photons.

[b]The beam had broad wings in HV mode, so only about one-fourth of the photons actually hit the target pixel, other photons were detected all over the detector.

detectors on the focal plane. The combined uncertainty from these two steps should be less than one hundred microns.

For planning the XRG scans, I carried out extensive simulations to determine the number of counts required per pixel, to attain the desired centroiding accuracy. I found that we can easily get this with as low as 360 counts per pixel, the centroid of pixels is determined to better than 8 μm (1-σ). We also need to map pixel boundaries reliably to measure the areas of pixels, as discussed later in Section 3.6. This places more stringent requirements on scans: the mean spacing between photon hits in a single pass of the X-ray beam should be $\sim$1 μm. Since the mean output count rate is 350–390 counts/s, this translates to a scan rate of 400μm s^{-1}, or about 8 hours per scan. For scheduling convenience, we extended the scan duration to 11 hours, such that the total duration of the start up procedure, LV scan, HV scan, and shutdown procedure is 24 hours.

After completing every scan, we ran several tests to verify integrity of the data. Parts of this work were completed as the SURF[3] project of Nancy Wu. The first check was to plot all stage positions to ensure coverage of the entire detector. Next, we checked the number of raw stage data packets in each pass of the stage over the detector and the duration of each pass took, looking for outliers. Nancy visually examined zoomed–in plots of each pass to verify that the translation stages were not stuck in any part of the scan. The stage hardware was used with only a small margin under the maximum load limits, so occasional malfunctions were detected in data. In only a few cases, we had to rerun scans due to hardware issues. In hindsight, we would have designed the setup with a higher safety margin, and compensated some of the stage loading by a pulley mechanism.

---

[3]Caltech Summer Undergraduate Research Fellowship, http://surf.caltech.edu/.

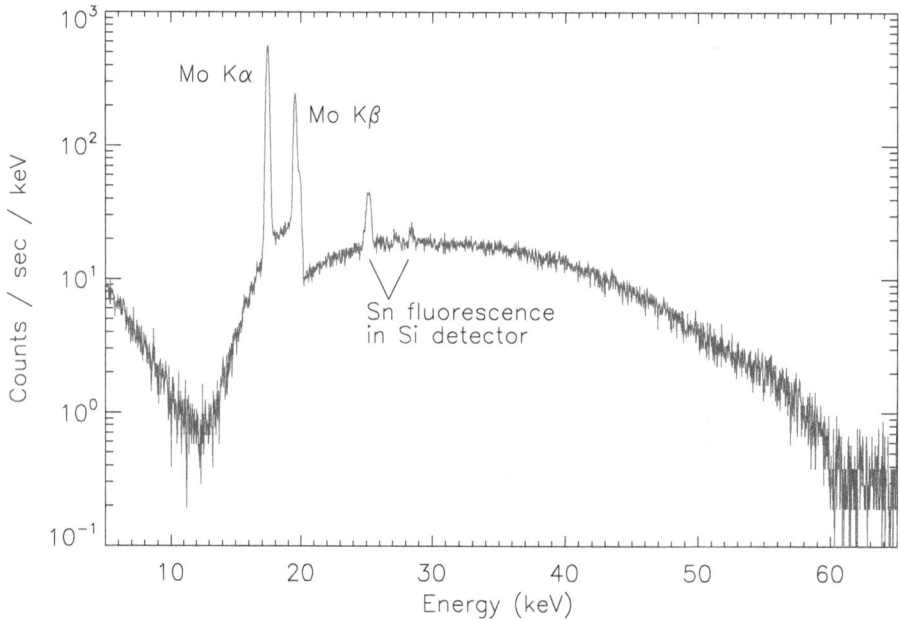

Figure 3.10. XRG spectrum in HV mode. The XRG was set at 60 kV, with a Mo target, and the spectrum was
         measured with a Si detector with a Cu + Sn collimator. The resultant spectrum is Bremsstrahlung
         superimposed with Mo Kα (17.4 keV), Kβ (19.6 keV), and the Mo K-edge (∼20 keV). To reduce the
         count rate, we added a 2.3 mm thick Al filter in the beam path. This preferentially absorbs the low
         energy photons, giving the dropoff around 15 keV.

3.5.2   Analysis and Results

Scan data recorded in a flight-like binary format are transferred to the NuSTAR workhorse linux server
"sarasvarti," where the files are converted to fits. All my analysis codes are written in IDL[4] and placed
under version control on a SVN server.

    The first analysis step is to calculate pixel centroids in stage coordinates. For events corresponding
to every (pixel, grade) combination, we want to calculate the centroid of that area of the detector which
generates such events. The net distribution of photons on the hybrid over a complete scan is uniform. So
the area centroids simply the mean position of all photons generating the particular (pixel, grade) triggers.
For LV scans, the beam is well contained within a narrow 50×70 μm spot, so the location of the translation
stages at the time of the event trigger is a good proxy actual photon interaction position. This average of
stage positions for all events forms the first estimate of the centroid. This position is iteratively refined
by rejecting any events outside a 5-pixel-wide box centered on the current centroid estimate. This process
is repeated until the centroid estimate shifts by less than 15 μm in an iteration, or up to 5 iterations.
Centroid calculations for HV scans are a bit more involved, as only about one-fourth of all photons are
in the core of the wing, with three-fourths being distributed in very broad wings extending beyond the

---
[4]http://www.exelisvis.com/idl/

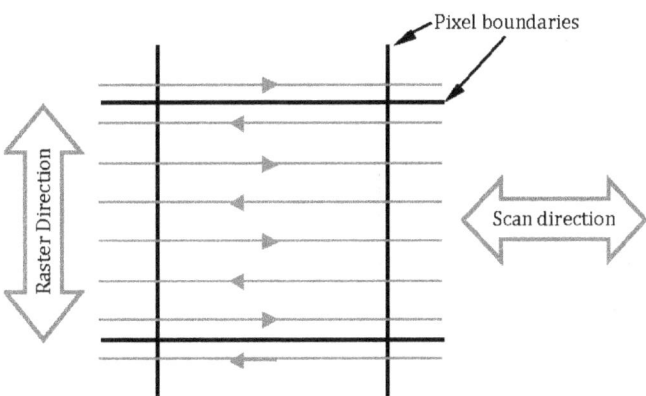

Figure 3.11. XRG scan procedure. The stages drive the hybrid so that the X-ray beam rasters over all pixels. Shown here, the beam traverses a pixel in the scan direction, then is stepped over by one beam width in the raster direction and traverses the pixel in the other direction. After completing such a scan over the entire detector, the scan and raster directions are swapped and the full scan is repeated.

hybrid. As a result, the mean position of all photon hits is always close to the center of the hybrid. To surpass this hurdle, we use LV centroids as the starting estimate. As before we select photons within a 5-pixel-wide box around this start estimate. Most of the photons in such a box are from the core of the beam, and the ~2% contribution from the wings can be ignored. The HV centroids are then iteratively refined as before.

The acid test for reliability of data is repeatability. I ran two complete LV calibration scans for H78, and analyzed the data sets independently to calculate centroids for each pixel. Comparing the results, we see that there is a small systematic offset between the centroids calculated from both scans: $\Delta X = 44$ μm and $\Delta Y = 2$ μm (Figure 3.12). After correcting for this offset, the scatter between the centroids calculated from both the runs were $\sigma_x = 11$ μm and $\sigma_Y = 8$ μm, comfortably smaller than the calibration requirements for NuSTAR.

Next, we use XRG scan data to calculate the boundaries and areas of individual pixels. The boundary mapping algorithm, developed as a part of Nancy Wu's SURF project, is as follows: the XRG scans are first split into individual passes over the detector. A typical flight calibration scan consists of about 400 passes in the Y direction and 300 passes in the X direction. For each of these passes, we assign events to the pixels with highest energy, independent of event grade. The boundary between two pixels is defined as the point where photons equally likely to trigger either pixel. XRG scan passes along the X direction are sensitive only to boundaries between pixel rows, while column boundaries are mapped from passes along the Y direction. The boundary of pixel $(C,\ R)$ is calculated as the intersection of the column C and row R using the IDL `inside`[5] routine. Areas are calculated for each pixel and uploaded to the SVN server.

3.6 Quantum Efficiency Measurements

Measurement of absolute QE of pixels is a two-step process: determining the "Bulk QE" of a part of the detector using radioactive sources, followed by measuring "relative QE" of various pixels from XRG scans.

---

[5]http://www.idlcoyote.com/tips/point_in_polygon.html

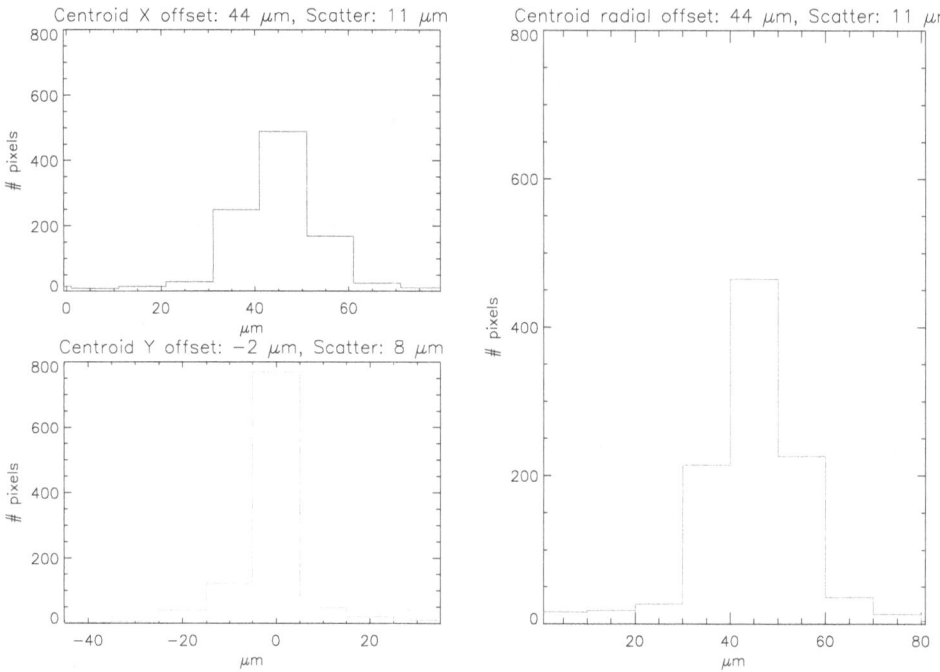

Figure 3.12. Comparison between centroids measured with two consecutive XRG LV scans for H78. The offsets were 44 μm in X and 2 μm in the Y direction. After correcting for this offset, the scatter between the centroids calculated from both the runs were 11 μm in the X direction and 8 μm in the Y direction.

3.6.1   Procedure

Absolute QE measurements were conducted with QE scans. We make minor modifications to the XRG scan setup: a well-calibrated radioactive source is inserted in the XRG beam tube and the fine slit is replaced by a QE mask (Section 3.3.1). This configuration is designed to illuminate a 12×12 region of the hybrid. In contrast to the continuous scanning procedure for XRG scans, the hybrid is parked at a single location for a "dwell" spanning 25–45 min depending on the radioactive source used. These durations are selected such that each dwell has $\gtrsim 10^5$ counts, decreasing the uncertainty from Poisson statistics to well under a percent. After each dwell, the StageGUI software moves the hybrid over by half a mask size and commences the next integration. Initial flight hybrid calibration scans consisted of a 6×6 grid of dwells. For dwells on the outer boundary of this grid, part of the source flux was incident beyond the active area of the hybrid, which rendered them less useful for absolute QE determination. For later calibration scans we narrowed down to a 4×4 grid of dwells, where all the source radiation was incident on an active part of the hybrid.

3.6.2   Analysis and Results

For the "Bulk QE" measurement, we split the scan data into individual dwells of the QE scan. Each dwell probes a different region of the hybrid. The data were reduced via standard NuSTAR pipelines to form spectra. We identified regions containing the photopeak and any tailing from a given line and added up all the counts in those regions to get count rates. The absolute QE is the ratio of the livetime corrected count rate for a hybrid to the total expected count rate from the radioactive source. For NuSTAR flight hybrids, the absolute QE is ~98% in a significant part of the energy range of interest (Figure 3.13). QE values for different regions of the detector are constant within measurement errors.

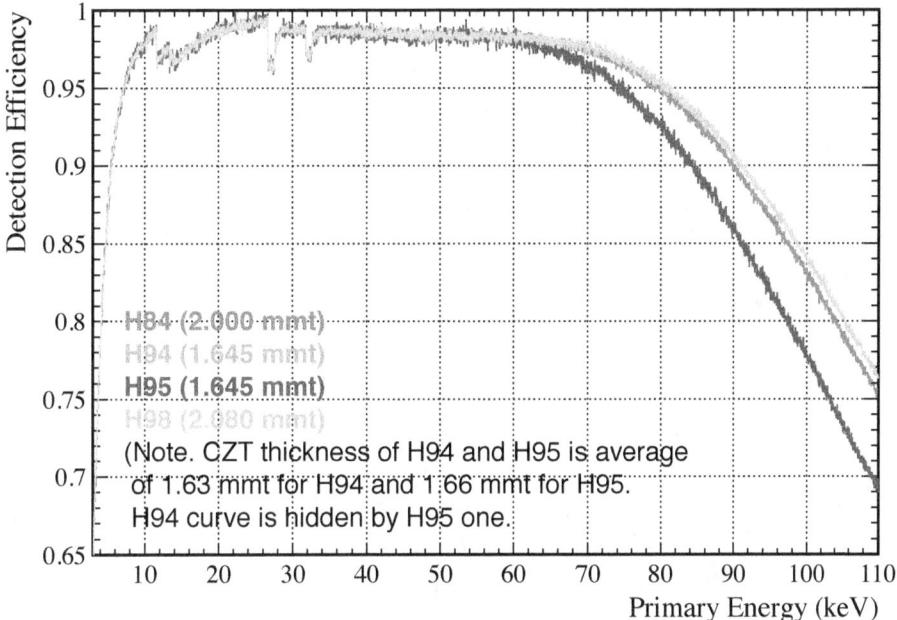

Figure 3.13. Absolute QE measurements for the four NuSTAR FPM-A hybrids. The QE is close to 98% over a significant part of the NuSTAR energy range. Plot courtesy Takao Kitaguchi.

To calculate absolute QE for individual pixels, we couple this data with "relative QE" measurements from XRG scans. While we have not calibrated the absolute flux of the XRG beam, we have verified that it is extremely stable over timescales of calibration scans (Section 3.4.2). Thus, any pixel-to-pixel variation in count rates in a XRG scan is indicative of the relative QE of pixels in that energy range. In particular, when the beam is near the center of some pixel, that pixel should detect all the counts from the beam, with no loss due to charge sharing or split pixel events. For each pixel, I extract counts from a 200×225 μm box around its grade 0 centroid (Section 3.5.2). Then I obtain the count rate by dividing by the hybrid livetime. A subtle but important effect arises from event triggers in other pixels. The data packet for each includes the livetime prior to the event. While events in other pixels should not be counted for measuring the count rate in the pixel of interest, we still have to add up the "prior" livetime from those events in the total livetime. The count rate is given by

$$
\begin{aligned}
\text{Count Rate} \;&=\; \frac{\text{Number of events in pixel of interest}}{\text{Total livetime of hybrid during the pass}} \\
&=\; \frac{\text{Number of events in pixel of interest}}{\Sigma\,(\textit{Prior}\ \text{time of events in ALL pixels during that pass})}.
\end{aligned} \tag{3.1}
$$

The normalized distribution of these count rates gives us the relative QE of pixels (Figure 3.14).

In default XRG LV calibration scans, typical pixels record between 1000 and 2000 counts in the central extraction region. Thus, even for a detector with uniform QE for all pixels, we expect a ∼3% scatter in the measured values of QE. As in Section 3.5.2, we tested the repeatability of relative QE measurements in laboratory data. I calculated relative QE for all pixels H78 pixels independently from two scans. Comparing the two sets of measurements, I found no evidence for statistically significant variations in QE among pixels. This is consistent with the observation that the bulk QE for various regions of the detector is constant within measurement uncertainties.

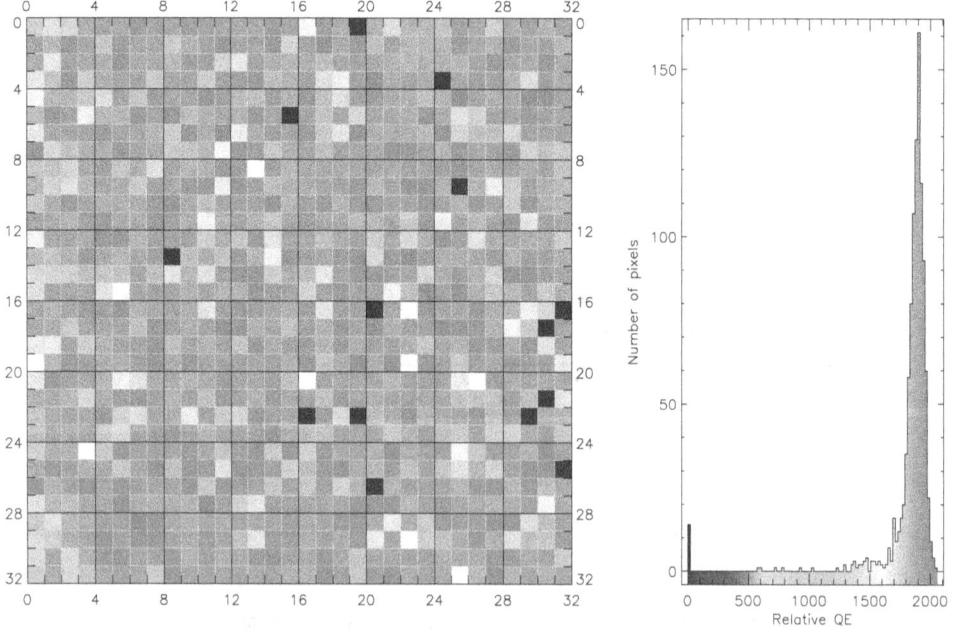

Livetime corrected count rate for that pixels events within 0.099900, 0.112490 mm of grade 0 centroids

Figure 3.14. Relative QE measurements for H82. The color scale and histogram denote the count rate in the centers of pixels measured in XRG LV scans, which is proportional to the pixel QE. Apart from the low QE pixels in the lower left, the scatter is within that expected from Poisson noise.

3.7 Transparency

As discussed at the start of this Chapter, we adapted the XRG lab setup to measure transparency of two components of NuSTAR: Be windows and optics thermal covers.

3.7.1 Beryllium Windows

The NuSTAR focal plane modules have Be windows in the photon path, to block off low energy photons (Figure 3.1). This is important as even photons with energies lower than the pixel trigger thresholds can increase the background noise. In extreme cases, a high flux of X-rays can slightly lower the gain of the detectors. The aim of laboratory calibrations was to measure the attenuation of Be windows as a function of energy and position on the ~3 inch diameter window.

To measure attenuation, we set up a radioactive source and a detector and measure the source flux (I_0) in some X-ray line. Then we insert the Be window between them and measure the attenuated line flux (I). The ratio I/I_0 gives the attenuation at that energy. We repeat this procedure using radioactive lines at various energies. The attenuation of the window depends on its thickness and composition. We can calculate the expected attenuation curve based on composition information provided by the vendor (Table 3.3). Then, we fit the curve to data with thickness as the only free parameter.

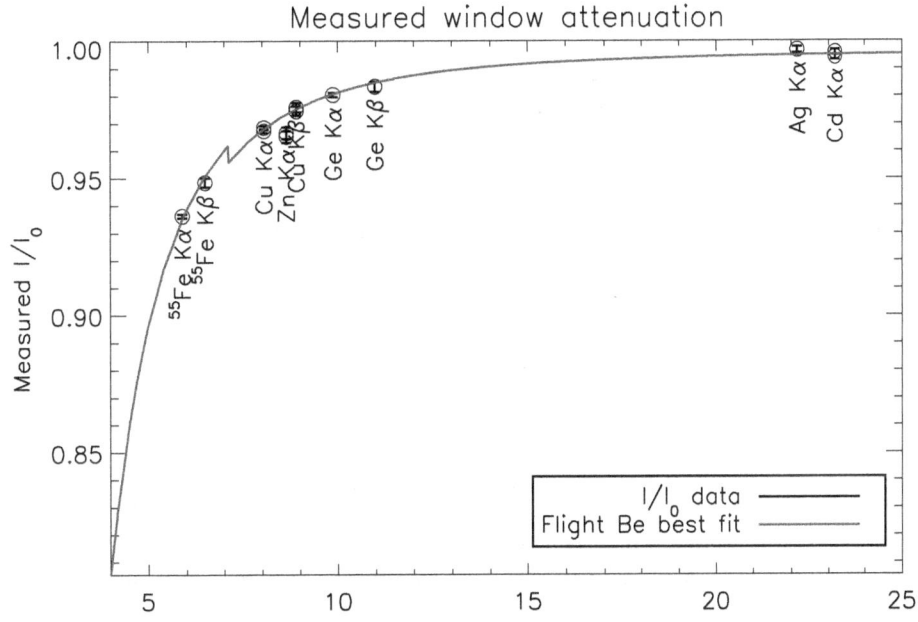

Figure 3.15. Best-fit attenuation curve for Be, using the composition from Table 3.3 and data from Table 3.4. We see that the vendor-specified composition gives a good fit to data.

The attenuation varies strongly with energy in the low energy (3–10 keV) range. At higher energies, the attenuation is lower and varies less strongly with energy—so this energy range can be sampled sparsely. We

Table 3.3. Be window composition

Element/Compound	Fraction by mass[a]
Be	0.99145
BeO	0.0070
Al	0.0004
C	0.0001
Fe	0.0008
Si	0.0002

[a]Composition based on information provided by Brush Wellman. Heat number 5226, Beryllium assay 99.4%

Table 3.4. Be window transmission

Line	Energy (keV)	Transmission
^{55}Fe Kα	5.89	0.9363(6)
^{55}Fe Kβ	6.49	0.9485(16)
Cu Kα	8.05	0.9685(7)
Zn Kα	8.64	0.9652(21)
Cu Kβ	8.91	0.9759(7)
Ge Kα	9.88	0.9805(7)
Ge Kβ	10.98	0.9834(18)
Ag Kα	22.16	0.9967(11)
Cd Kα	23.17	0.9941(9)

use the ^{55}Fe Kα and Kβ lines as the lowest energy probes. For intermediate energies, we use fluorescence lines from various materials irradiated by the XRG (Table 3.4). We modified the XRG setup to mount a target at 45° to the X-ray beam and placed the Amptek Si detector to intercept fluorescent X-rays. The observed spectrum consists of strong fluorescence lines superposed on a relatively weak compton scattered XRG spectrum.

We undertook the following measurements for 3 Be windows: first, we measured the full attenuation curve at the center of the window with I/I_0 measurements. Next, we move the window and measure the attenuation at a single energy at various positions on the window. The attenuation for each of the three windows is in excellent agreement with expectations. The window thickness varies slightly with radius (within manufacturing specifications). On examining the results, the more uniform windows were selected as flight windows. These windows are nominally 110 μm thick and one of them shows a slight radial increase in attenuation (Figure 3.16).

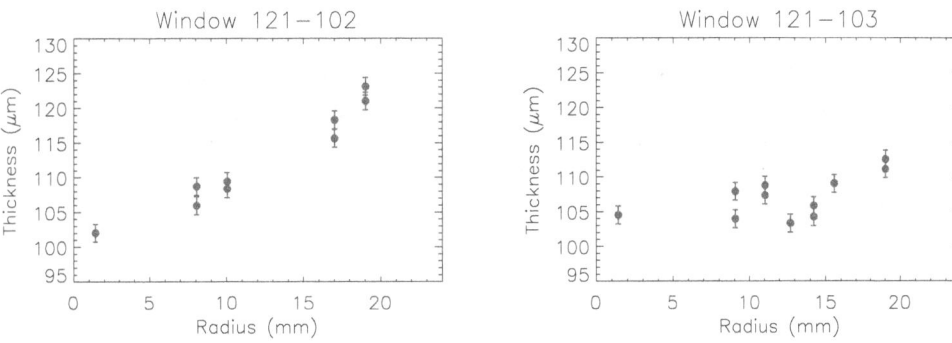

Figure 3.16. Radial variation of thickness for flight Be windows. The thickness was calculated using the 5.89 keV ^{55}Fe Kα and 6.49 keV ^{55}Fe Kβ lines.

3.7.2   Optics Cover

Thermal covers for optics are made of Mylar film. Ribs added for structural integrity occupy about 1% of the area of the cover. Calibration requirements for these covers are folded into requirements for overall optics calibrations. We were tasked with verifying the transparency of these covers at low energies. We selected the ^{55}Fe source and the Amptex Si detector for calibration. The source radiates a rather wide X-ray beam, while the detector has a 2.38 mm circular collimator. As in the Be window calibrations, we first lined up the source and detector and measured the source flux I_0. We mounted a "flight spare" optics cover close to the detector and measured the attenuation for at various points on the surface (Figure 3.17). The mean transparency at 5.9 keV is 96%, in exact agreement with design requirements.

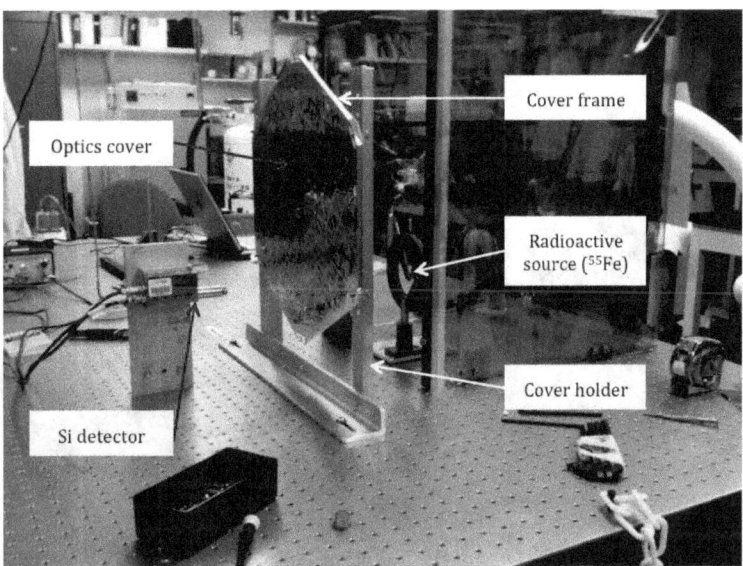

Figure 3.17. Setup used for measuring the transparency of the optics thermal cover at low (∼6 keV) energies.

3.8   Summary

As I write this thesis, all the calibrated components have been mounted in the Focal Plane Modules, and the instrument and spacecraft are ready for launch. Calibration data has been processed into data products required for analyzing astrophysical source data from NuSTAR. The results are being converted into the appropriate format for distribution with NuSTARDAS. With NuSTAR scheduled for launch in Summer 2012, we look forward to the exciting science to come!

Part Two: Masses of Neutron Stars

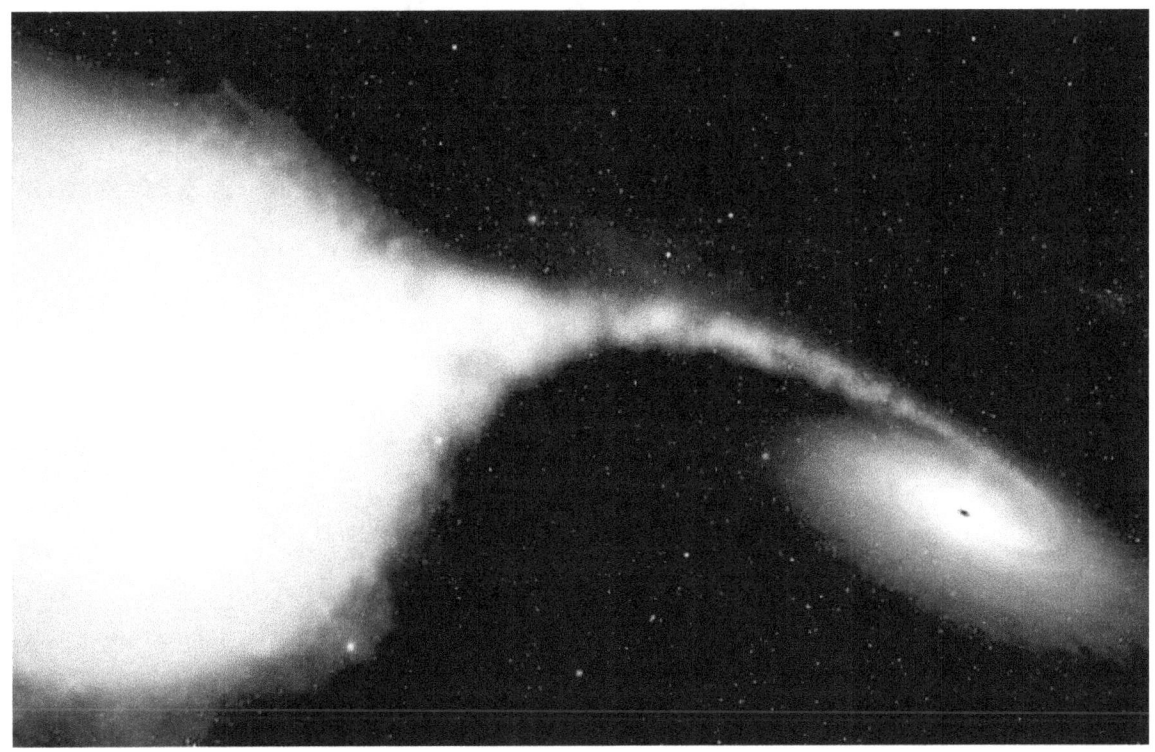

X-Mas: A Search for
eXtra Massive Neutron Stars

Image: Artists illustration of a galactic Black Hole high-mass X-ray binary Cygnus X-1.
Credit: ESA/Hubble, via Wikimedia Commons.

Chapter 4

High-Mass X-ray Binaries

This part of the thesis focuses on the masses of neutron stars. In this Chapter, I motivate the case for measuring neutron star masses. Then I narrow down the sample to high-mass X-ray binaries and discuss the methods and techniques for measuring masses. In Chapters 5 and 6, I provide the results of X-Mas: our systematic survey of neutron star masses in high-mass X-ray binaries to shed light on the formation processes of NSs. In Chapter 7, I describe some constraints on the evolution of a two solar mass neutron star.

4.1 Weighing in on Neutron Stars

Neutron stars (NSs) are compact remnants of high mass stars. The following three questions are central to our study of neutron stars: (1) What is the relation between the mass of the progenitor star and the mass of the neutron star (the initial-final mass mapping)? (2) Are there multiple channels to form neutron stars? (3) Are there exotic states of matter in the interiors of neutron stars? As explained below, measuring the masses of neutron stars helps answer these three questions.

The global structure of a neutron star depends on the equation of state (EOS) of matter under extreme conditions, i.e., the relation between pressure and density in the neutron star. Given an EOS, the maximum mass of a neutron star can be calculated (Figure 4.1). If the EOS is "Soft," then kaon condensates or strange matter can form in the interior of neutron stars, and an upper limit of <1.55 $M_\odot$ is expected for NS masses (Lattimer & Prakash, 2005). Our current understanding of nuclear physics predicts a "stiff" EOS, where a given density can support higher pressure. This claim is further bolstered by the discovery of a 2 $M_\odot$ NS (Demorest et al., 2010; Lattimer et al., 2010).

The primary observables of neutron stars—spin periods, radii, and masses—can be used to constrain their properties and internal structure. The fastest pulsar periods and periods of quasi-periodic oscillations in accreting NSs provide limits on the radii and masses of neutron stars (Lattimer & Prakash, 2007). Radii inferred from neutron star cooling or modeling of photospheric expansion bursts have also been used to test NS models from various equations of state (Steiner et al., 2010). Dynamical measurements of masses of NS in binaries provide an important test to these models. In compact object binaries (NS–NS, NS–white dwarf), very precise mass measurements can be obtained from relativistic effects like the advance of periastron (Freire et al., 2008) or measurement of the Shapiro delay (Demorest et al., 2010).

The mass of the supernova remnant depends on the evolutionary state of the progenitor, the nature of core collapse (CC), and post-CC evolution. I restrict the discussion here to birth masses of neutron stars, and revisit some aspects of post-CC evolution in our work on the companion to the 2 $M_\odot$ PSR J1614-2230

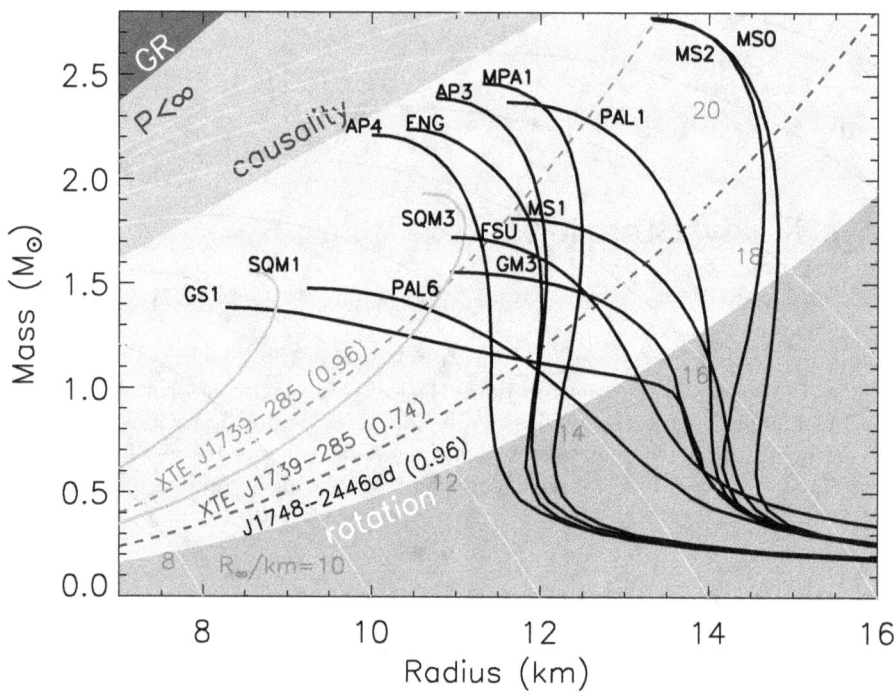

Figure 4.1. Theoretical mass-radius relationships for neutron stars. Solid curves are models calculated assuming different equations of state (EOS), labeled as per Lattimer & Prakash (2007). The discovery of massive neutron stars, e.g., the 2 $M_\odot$ PSR J1614-2230, rules out the soft equations of state. Figure credit (Lattimer & Prakash, 2007), reproduced with permission from Elsevier.

in Chapter 7. Model calculations by Timmes et al. (1996) predict that type II (core-collapse) supernovae should form NSs with a bimodal distribution of masses peaked at 1.28 and 1.73 $M_\odot$, while Ib supernovae will produce NSs with masses in a small range around 1.32 $M_\odot$. If the NS is born in a binary, then it may accrete matter from the companion, spreading out these distributions. In addition, neutron stars formed in electron-capture supernovae likely have lower mass (Nomoto, 1984), and may be present in some of the binary pulsars (Schwab et al., 2010). Knigge et al. (2011) suggested that this alternative formation channel might also account for the observed bimodality in the properties of Be X-ray binaries, where in contrast to the majority of systems, some have low eccentricity, suggesting a small natal kick.

  As of summer 2012, masses have been measured for over fifty five neutron stars.[1] The dynamical masses of neutron stars can be measured only in binary systems. Most of the mass measurements are for NSs in radio pulsar binaries or NS-white dwarf (WD) binaries. Several groups have analyzed the mass measurements and inferred that NS masses have a bimodal distribution (Kiziltan et al., 2010; Zhang et al., 2011; Valentim et al., 2011; Özel et al., 2012). This bimodality likely stems from NSs accreting mass from

---

[1]A complete list of measured masses is at http://stellarcollapse.org/nsmasses.

their companion during the spin-up (recycling) process. The differences in mass accreted in the recycling process may dominate over any intrinsic variations in the birth masses of such pulsars. Being restricted to these two types of binaries may not reveal the complete range of neutron star masses.

The third major class of NS binaries, X-ray binaries (XRBs), trace distinct evolutionary pathways for NSs. XRBs are sub-divided into high-mass X-ray binaries (HMXBs) and low mass X-ray binaries (LMXBs) based on the spectral type and mass of the companion star. My work focuses on the HMXB subset of this population, which we explore in detail next.

4.2   High-Mass X-ray Binaries

High mass X-ray binaries (HMXBs) are binary systems containing a neutron star or a black hole (secondary) and massive ($\gtrsim 8$ M$_\odot$) OB (primary) companions, with orbital periods ranging from days to months. In 2006, there were 114 known HMXBs in the Milky Way (Liu et al., 2006) and 128 in the Magellanic Clouds (Liu et al., 2005). A new population of highly obscured HMXBs was discovered by the *Integral* satellite (Chaty et al., 2010). In this thesis, I discuss HMXBs with a NS as the secondary, unless explicitly stated otherwise.

High mass X-ray binaries can be observationally divided into several types:

1. *Be X-ray binaries* have a non-supergiant, fast-rotating Be star as the optical companion (e.g., Cep X-4, GX 304 − 1). Compared to normal B-type stars, Be stars show emission lines and excess infrared emission at some point in their lives. These features are attributed to the formation of an equatorial disc around the star, the origins of which are not fully understood yet (Reig, 2011, and references therein). Most Be XRBs have eccentric orbits ($e \gtrsim 0.3$) and show transient X-ray emission near periastron passages, though lower luminosity persistent systems also exist. Almost all Be XRBs show X-ray pulsations and are inferred to have NS counterparts: there are no candidate Be–black hole binaries (Paul & Naik, 2011; Reig, 2011).

2. *Classical X-ray binaries* are XRBs where the optical companion is a supergiant OB star (Chaty, 2011). This companion emits a substantial wind, and the NS orbits inside this wind (e.g., Vela X-1). Although capture from a high-velocity stellar wind is inefficient, the large mass-loss rate in the wind can result in an appreciable mass accretion rate onto the neutron star that is sufficient to emit radiation in X-ray band (Paul & Naik, 2011). In some objects of this type, the orbit is compact enough that the OB star fills its Roche lobe. For such objects, the accretion rate is greatly enhanced and may be mediated through an accretion disc (e.g., SMC X-1, LMC X-4).

3. *Supergiant Fast X-ray Transients (SFXTs)* are hard X-ray transients with a high dynamic range ($10^3 − 10^5$) in their X-ray lightcurves (e.g., IGR J17544−2619, AX J1841.0−0536). They show recurrent outbursts on few-hour timescales, superposed on outbursts lasting a few days (Sidoli, 2011). This is a relatively new class with about 10 members, all of which are associated with blue supergiants. Pulsations have been detected in only five of the SFXTs (Sidoli, 2011), but it is generally assumed that all SFXTs harbor neutron stars. This class is sometimes considered a subclass of classical HMXBs and it is possible that SFXTs and classical wind-fed HMXBs lie along a continuum (Chaty, 2011).

NS masses in HMXBs will be close to the birth masses. After the neutron star is formed, it can accrete mass from the companion by various mechanisms like wind accretion, disk accretion, or Roche lobe

Table 4.1.   Masses of neutron stars in high-mass X-ray binaries

Object	M_X	M_{opt}	Reference
4U1700-37	2.44 ± 0.27	58 ± 11	Clark et al. (2002)
4U1538-52	$1.06^{+0.41}_{-0.34}$	$\cdots$	van Kerkwijk et al. (1995)
SMC X-1	$1.06^{+0.11}_{-0.10}$	$15.7^{+1.5}_{-1.4}$	van der Meer et al. (2007)
Cen X-3	$1.34^{+0.16}_{-0.14}$	$20.2^{+1.8}_{-1.5}$	van der Meer et al. (2007)
LMC X-4	$1.25^{+0.11}_{-0.10}$	$14.5^{+1.1}_{-1.0}$	van der Meer et al. (2007)
Vela X-1	1.88 ± 0.13[a]	23.1 ± 0.2	Barziv et al. (2001); Quaintrell et al. (2003)
	2.27 ± 0.17[b]	27.9 ± 1.3	
EXO 1722-363	1.4 ± 0.4[a]	13.6 ± 1.6	Mason et al. (2010)
	1.5 ± 0.4[b]	15.2 ± 1.9	
IGR J18027-2016	1.4 ± 0.2[a]	18.6 ± 0.8	Mason et al. (2011b)
	1.6 ± 0.3[b]	21.8 ± 2.4	
OAO 1657-415	1.4 ± 0.3[a]	14.2 ± 0.4	Mason et al. (2011a)
	1.7 ± 0.3[b]	17.0 ± 0.7	

[a] Assuming edge-on orbit ($i = 90°$).

[b] Assuming Roche lobe filling companion ($\beta = 1$).

overflow. If the accretion is spherically symmetric, it will be limited by the Eddington accretion rate. For a NS of radius ~ 14 km, the limiting accretion rate is $\dot{M}_{Edd} \approx 2 \times 10^{-8}$ $M_\odot$ yr^{-1}, independent of its mass. In this case, in the $\sim 10^7$ yr lifetime of the OB companion, the NS can accrete at most ~ 0.1 $M_\odot$. Hence, measuring the masses of NSs in HMXBs allows us to measure the distribution of their birth masses.

Radial velocity measurements for these OB stars are complicated by several factors. Some of the spectral lines, especially Hα and Hβ, are variable. In some objects, lines are produced at different parts of the stellar photosphere and may have different radial velocities. Many of the systems are highly extincted and are visible only in IR. However, among the very few measurements, there is a large scatter in masses (Table 4.1). Among the five HMXBs with reasonably secure masses, one has a high value of $M = 1.8 \pm 0.3$ $M_\odot$ (Vela X-1; Barziv et al., 2001; Quaintrell et al., 2003). As discussed above this indicates that this neutron star may have been born heavy.

4.3   NS mass measurements in HMXBs

Motivated by the spread in known NS masses in HMXBs despite them being close to their masses at birth, we undertook "X-Mas," an extensive program to measure masses for more such neutron stars by radial velocity (RV) studies of the OB companions. Here I outline the general method, observations and data analysis for such mass measurements. In Chapter 5, I present first results for a few HMXBs obtained from data taken with the 200" Hale telescope at Palomar, followed by the constraints on the mass of the compact object in an eclipsing HMXB in M33 (Chapter 6).

4.3.1 Method

The masses of both components in a binary can be calculated by characterizing its orbit. A binary orbit is fully described by seven orbital elements: the semi-major axis (a), period (P), eccentricity (e), inclination (i), longitude of the ascending node (Ω), longitude of periastron (ω) and the time of periastron passage (T_0) (Figure 4.2). In HMXBs, the orbital elements for the NS are often calculated by X-ray pulse timing measurements in different parts the orbit. In our study, we measure the radial velocity of the OB companion around the orbit. Using the optical and X-ray data, we can calculate the mass function, and in some cases the mass of the neutron star.

We follow the method introduced by Joss & Rappaport (1984), which is widely used (see for example, van der Meer et al., 2007; Mason et al., 2011b). We can write the masses of the optical companion and the X-ray source (M_{opt} and M_{X}, respectively) in terms of the mass functions:

$$M_{\mathrm{opt}} = \frac{K_{\mathrm{X}}^3 P (1-e^2)^{3/2}}{2\pi G \sin^3 i}(1+q)^2, \tag{4.1}$$

and

$$M_{\mathrm{X}} = \frac{K_{\mathrm{opt}}^3 P (1-e^2)^{3/2}}{2\pi G \sin^3 i}\left(1+\frac{1}{q}\right)^2, \tag{4.2}$$

where K_{X} and K_{opt} are the semiamplitudes of the respective radial velocity curves, (i) is the inclination,[2] and q is the mass ratio of the components, defined as

$$q \equiv \frac{M_{\mathrm{X}}}{M_{\mathrm{opt}}} = \frac{K_{\mathrm{opt}}}{K_{\mathrm{X}}}. \tag{4.3}$$

In practice, X-ray timing results are often quoted as the projected semi-major axis of the pulsar's orbit in light-seconds, $a_{\mathrm{X}} \sin i$, from which we get $K_{\mathrm{X}} = 2\pi c\, a_{\mathrm{X}} \sin i / P$. Our optical or IR spectra can provide a value for K_{opt}.

The orbital inclination is usually unknown, making it impossible to solve explicitly for component masses. Equations (4.1) and (4.2) are then recast in terms of the mass function f:

$$
\begin{aligned}
f_{\mathrm{X}} &\equiv \frac{M_{\mathrm{X}}^3 \sin^3 i}{(M_{\mathrm{X}} + M_{\mathrm{opt}})^2} \\
&= \frac{K_{\mathrm{opt}}^3 P (1-e^2)^{3/2}}{2\pi G} \\
&= 1.036 \times 10^{-7}(1-e^2)^{3/2} K_{\mathrm{opt}}^3 P\ \mathrm{M_\odot},
\end{aligned} \tag{4.4}
$$

where mass is measured in solar masses, velocity in kilometers per second and period in days.

Since $\sin i \leq 1$, we can use Equation (4.4) to calculate a lower limit M_{X} as a function of M_{opt}. In turn, M_{opt} can be estimated from the spectral class of the OB star. For most of the systems in our sample, we calculate such lower bounds on masses (Chapter 5).

If the HMXB shows eclipses, we can calculate the inclination from the eclipse duration. Assuming a spherical companion star, the inclination of the system is related to the eclipse half-angle[3] θ_e, the stellar

---

[2] i is defined as the angle between the orbital plane to the line of sight, with $90°$ being edge-on.

[3] The eclipse half-angle θ_e, or more specifically the semi-eclipse angle of the neutron star, represents half of the eclipse duration.

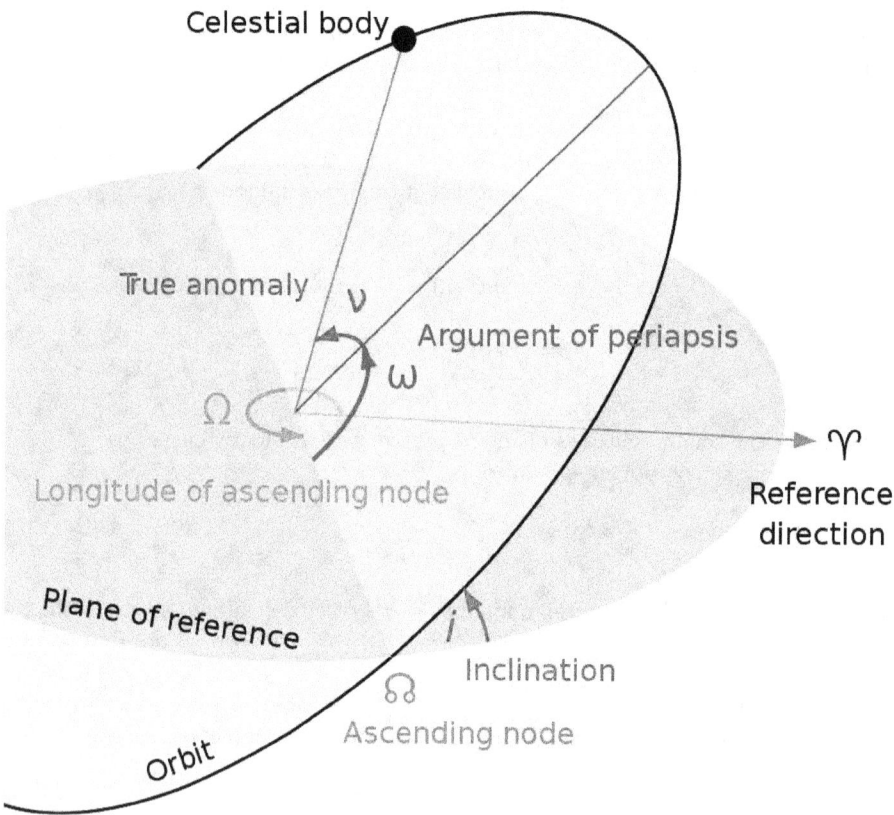

Figure 4.2. The elements of an elliptical orbit. The observer is located below the the gray horizontal plane of reference, viz., the sky plane. All angles are measured with the center of mass of the orbit as the vertex. The inclination of the orbit to the sky plane (i) is measured such that $i = 90°$ for an edge-on orbit. The ascending node is where the object crosses the plane of the sky, receding from the observer. The longitude of the ascending node (Ω) is the angle between the reference direction Υ (north) and the ascending node. It can be measured only for resolved binaries, and does not enter our calculations. The argument of periastron (ω) is the angle in the plane of the orbit between the periastron (periapsis) and the ascending node. At a given time T, the location of the object along the orbit is given by its true anomaly ν. The reference time T_0 is usually defined as the time when the object is at periastron. Image reproduced from Wikimedia Commons (http://commons.wikimedia.org/wiki/File:Orbit1.svg).

radius R, and the orbital separation a by

$$\sin i = \frac{\sqrt{1 - (R/a)^2}}{\cos \theta_e}. \tag{4.5}$$

Following the approach in Rappaport & Joss (1983), the radius of the companion star is some fraction of the effective Roche lobe radius,

$$R = \beta R_L, \tag{4.6}$$

where R_L is the sphere-equivalent Roche lobe radius. We will refer to the fraction β as the "Roche lobe filling factor." Combining Equations (4.5) and (4.6) yields

$$\sin i = \frac{\sqrt{1 - \beta^2 (R_L/a)^2}}{\cos \theta_e}. \tag{4.7}$$

For a co-rotating secondary, Eggleton (1983) gives an expression for R_L/a, the ratio of the effective Roche lobe radius and the orbital separation:

$$\frac{R_L}{a} \approx \frac{0.49 q^{-1/3}}{0.6 q^{-2/3} + \ln(1 + q^{-1/3})}. \tag{4.8}$$

In some cases, the secondary star might not be corotating with the system. The ratio of the rotational frequency of the optical companion to the orbital frequency of the system is defined as Ω. In other words, it is a measure of the degree of synchronous rotation, where $\Omega = 1$ is defined to be synchronous. In this case, we use approximations by Rappaport & Joss (1983):

$$\frac{R_L}{a} \approx A + B \log q + C \log^2 q, \tag{4.9}$$

where the constants A, B, and C are

$$\begin{aligned} A &= 0.398 - 0.026\Omega^2 + 0.004\Omega^3, \\ B &= -0.264 + 0.052\Omega^2 - 0.015\Omega^3, \\ C &= -0.023 - 0.005\Omega^2. \end{aligned} \tag{4.10}$$

These four expressions give the value of R_L to an accuracy of about 2% over the ranges of $0 \le \Omega \le 2$ and $0.02 \le q \le 1$ (Joss & Rappaport, 1984). If we substitute $\Omega = 1$ in Equation (4.10), then the R_L/a values of Equations (4.8) and (4.9) agree to within 1% in the range $0.01 \lesssim q \lesssim 1$.

In noneclipsing binaries, $\sin i$ is usually unconstrained. For such binaries, we can only place lower limits on M_X (Equation (4.4)). In binaries without accurate NS pulse timing data, neither K_X nor $a_X \sin i$ are known. For such systems, the mass estimates are indirect. Where good spectra are available, they can be used to estimate the spectral type and thereby the mass of the optical component. In rare cases where the distance to the binary is known, it provides an independent constraint on the physical scale of the system—for example by calculating the absolute magnitude of the components (Chapter 6). However, calculating masses from these methods is model dependent to some extent.

4.3.2 Observations

The accurate measurement of radial velocities, and in turn of orbital solutions, depends on several factors before, during and after the observing run. This chapter provides a general overview of the observing procedure, more details of specific observing runs are given in Chapters 5 and 6.

For our program, the key factors for an observing run are

1. Select proper instrument settings to cover all prominent spectral features with sufficient wavelength resolution,
2. Calculate the desired SNR and deduce the exposure time,
3. Select epochs of observation to maximize phase coverage,

Depending on the spectral and luminosity class, OB star spectra have strong Balmer series lines, as well as lines from He, C, N, O etc. in the blue part of the optical spectrum ($\sim$3500 Å–5500 Å). In contrast, there are considerably fewer spectral features at wavelengths longer than $\sim$5500 Å. The SNR of the spectra at these wavelengths is lower, as the flux of OB stars in this wavelength range is lower than the blue range, and the background noise is higher due to intrinsic sky emission. Hence, we concentrate on the blue spectra for measuring radial velocities. The absorption lines are rotationally broadened to few hundred kilometers per second, corresponding to a FWHM of few angstroms in this wavelength range. Thus, low resolution spectrographs with resolving power of few thousand are sufficient to resolve the lines. Considering the large desired wavelength coverage, we selected the Double Beam Spectrograph on the 5 m Hale telescope at Palomar (DBSP; Oke & Gunn, 1982) for most of our observations. For the fainter target XMMU J013236.7+303228 (Chapter 6), we obtained data with the Low Resolution Imaging Spectrograph on the 10 m Keck-I telescope (LRIS; Oke et al., 1995), with upgraded blue (McCarthy et al., 1998; Steidel et al., 2004) and red cameras (Rockosi et al., 2010).

We planned exposures to ensure all acquired spectra have a high enough signal-to-noise ratio for measuring radial velocities with desired accuracy, as follows: From Equation (4.1), the mass depends on K_{opt}^3. To measure M_{opt} with $\sim$10% accuracy, we need to measure K_{opt} with $\sim$3% accuracy. Typical values of K_{opt} for HMXBs are few tens of kilometers per second, and we had planned a few epochs per target, so this translates to a requirement of measuring velocity with an accuracy of about 5 km s^{-1} per epoch. I simulated DBSP and LRIS spectra and measured radial velocities from them to calculate expected errors as a function of SNR. Our project requires SNR of $\sim$100 per pixel, which translates to an exposure time of 80 min for a $m_V = 15$ target with DBSP. In practice, we divided this exposure time into four or more spectra, measured velocities on each spectrum independently, and combined the results into a single velocity measurement for that epoch.

To fit an orbit to radial velocity data, measurements need to be well spaced in orbital phase. Ideally, monitoring programs like RV measurements are best executed on queue–scheduled telescopes. For our program at Palomar, we applied for well-spaced nights. We obtained good phase coverage for most of our targets from pseudorandom spacing of these observations. Our target for the Keck observing run (XMMU J013236.7+303228) has a period of 1.73 days, so we obtained observations on two full consecutive nights. Unfortunately, the phase coverage was not optimized.

Last, but not the least, we ensured careful logging of our observing runs. Any anomalies like cloud cover, worsened seeing, tracking issues, etc., can distort the spectrum. Some of these effects can be canceled out in data analysis, while others cannot. While analyzing data, we reject some spectra based on observing logs before calculating the final orbit (see, for example, Chapter 6). In this regard, the equatorial mounting of the 200" Hale telescope is a great advantage, as it gives excellent tracking performance even at high airmass (low altitude).

4.3.3 Data Reduction and Analysis

After the observing runs, RV measurements can broadly be broken into two steps. The first step, data reduction, refers to converting the raw telescope data into spectra: a table of flux as a function of wavelength. The second step, analysis, involves calculating stellar parameters, radial velocities and binary orbits from the reduced spectra.

I used IRAF[4] to reduce all spectra. I trim and bias-subtract all spectra, followed by a spectral response correction to remove the small-scale variations in raw data. Then I use arc lamp exposures acquired at the start of the observing night to calculate the wavelength solution and extract 1-D spectra from raw data. The standard procedure often involves using arc lamp exposures taken closest to the science exposure. However, I have verified that for both DBSP and LRIS, the wavelength solution is extremely stable modulo a constant offset. In other words, the relative spacing of various lines in the arc lamps is constant in all exposures taken at various telescope pointings through the night. Emission lines in the sky spectrum are extremely stable (Figueira et al., 2010) and provide an excellent reference for measuring and correcting for this offset. I use the O I 5577.340 Å line to calculate wavelength offsets for LRIS spectra, and the Hg 4046.564 Å, 4358.336 Å lines for DBSP spectra. Sesar et al. (2012, in prep) adopted this approach, and found that the velocity measurements using this method are accurate to $\sim$2.5 km s^{-1}. Finally, I use the standard and calibrate procedures in IRAF to remove large-scale variations and flux calibrate the spectra.

I have developed a software (getvel) in IDL to analyze these fluxed spectra and measure radial velocities. One approach to measuring radial velocity from a spectrum is to measure the location of individual spectral lines, and to average that information (van der Meer et al., 2007). However, we often see that lines like Si IV, C III that show up in combined spectra are too weak to be detected in individual spectra. In order to utilize this information buried in noise, we measure radial velocities by fitting the entire spectrum with a template. I first determine the spectral type of the target by comparing the lines with the Gray spectral atlas[5] and those referred to in Walborn & Fitzpatrick (1990). Then I select a close range of model stellar spectra (templates) from the GAIA spectral library by Munari et al. (2005). For each observed spectrum, I measure the seeing using the width of the spectral trace in the slit direction. Then I generate an instrument response function by taking a Gaussian matched to the seeing, truncating it at the slit size, and convolving it with the pixel size. I convolve the fluxed templates this instrument response, then redshift them to a test velocity. If the extinction is known, I redden the spectrum using coefficients from Cox (2000). I use IDL[6] mpfit (Markwardt, 2009) to calculate the normalization to match this spectrum with the observed spectrum, and measure the χ^2. By minimizing the χ^2 over test velocities, I find the best-fit velocity and the error bars. Lastly, I convert this velocity to a barycentric radial velocity using the baryvel routine in Astrolib (Landsman, 1993).

I use the measured radial velocities and known binary parameters like the period P, phase T_0, and eccentricity e to calculate the best-fit orbital solution for the secondary star. Using the orbital solution, primarily K_{opt}, I calculate the mass of the neutron star following the procedures in Section 4.3.1.

---

[4]http://iraf.noao.edu/
[5]http://ned.ipac.caltech.edu/level5/Gray/frames.html
[6]http://www.ittvis.com/ProductServices/IDL.aspx

Chapter 5

X-Mas at Palomar

"X-Mas" at Palomar is a systematic radial velocity survey of HMXBs using the 200" Hale telescope, with the aim of understanding the distribution of masses of NS in HMXBs. We obtained data with the Double Beam Spectrograph on six nights in 2009, hereafter referred to as "Epochs" (Table 5.1). We used the D55 dichroic to split incoming light into the blue and red channels at 5500 Å. On the blue side, we used a 1200 lines/mm grating blazed at 5000 Å, with a resolution of 0.55Å/pix. On the red side we used the 1200 lines/mm grating blazed at 7100 Å, with a resolution 0.65 Å/pix. Depending on the observing conditions, we used 1″ or 1″.5 wide slits. Further details of planning the observing runs and data analysis steps are detailed in Section 4.3.

We selected a total of eight NS–HMXBs with known orbital periods. The nature of the compact object can be inferred from the X-ray spectrum, as accreting black holes have harder spectra than NSs. In some objects, the presence of a NS is confirmed by detection of X-ray pulsations. Here, I present the RV measurements and analysis of six HMXBs. We obtained orbital solutions and calculate masses for two of these objects: IGR J17544-2619 (Section 5.1) and SAX J2103.5+4545 (Section 5.2). Our RV data for the Be HMXB GRO J2058+42 can used to calculate the mass if more X-ray timing data is obtained (Section 5.4). Our velocity measurements for 1H 2138+579 do not show a clear orbital trend, which is discussed in Section 5.3. We are in the process of calculating orbital solutions for two more targets: 4U 2206+543 and KS 1947+300. After we completed observations, it was discovered that two candidates were misidentified. Negueruela & Schurch (2007) had proposed a tentative counterpart for AX J1820.5−1434, however a later study by Kaur et al. (2010) confirmed another star as the optical component of this HMXB. In case of IGR J17391-3021, the counterpart detected in IR (Smith, 2004) was misidentified in the optical band.

In this chapter I provide the early results from "X-Mas at Palomar," adding two crucial data points to the small sample of NS with known masses in HMXBs.

5.1   IGR J17544-2619

IGR J17544-2619 was discovered by *Integral* during an outburst in 2003 (Sunyaev et al., 2003). The source is a SFXT consisting of a neutron star (in 't Zand, 2005) and a supergiant O-type primary (Smith, 2004; Pellizza et al., 2006). Follow-up observations with RXTE (Rossi X-ray Timing Explorer) established the spin and orbital period of the NS: $P_{\rm spin} = 71.49$ s, $P_{\rm orb} = 4.9278(2)$ d (Drave et al., 2012). However, the pulsations were seen only in one observations and could not be reproduced in our analysis of the same data. This precludes the inference of other orbital parameters, e.g., $a_{\rm X} \sin i$.

The supergiant primary is a O9 Ia star, with $T = 31000$ K, and has foreground extinction $A_V = 6.3 \pm 0.4$ mag (Pellizza et al., 2006). We measure the temperature and extinction by fitting our spectra

Table 5.1.   DBSP observations

Label	Epoch[a]	Seeing[b]	Barycentric radial velocity (km s^{-1})			
			IGR J17544-2619	SAX J2103.5+4545	1H 2138+579	GRO J2058+42
a	2009 June 19	1.9″	-9 ± 3	-63 ± 2	-83 ± 6	-82 ± 4
b	2009 June 29	1.9″	-12 ± 4	-72 ± 2	-125 ± 8	-90 ± 5
c	2009 July 31	1.9″	-46 ± 4	-51 ± 2	-105 ± 7	-102 ± 4
d	2009 August 16	2.0″	-42 ± 3	-85 ± 3	-91 ± 8	-69 ± 5
e	2009 August 23	1.8″	-1 ± 3	-65 ± 4	-112 ± 7	-103 ± 3
f	2009 September 18	2.0″	$\cdots$	-68 ± 3	-93 ± 13	-108 ± 5

[a]Dates in Pacific time at the start of an observing night.

[b]Seeing is measured by fitting a Gaussian to the trace of the blue spectrum in the spatial direction. The number quoted here is the median value for all science exposures for that night.

with model spectra with solar metallicity, no rotational broadening, and surface gravity $\log(g) = 4.0$ (Munari et al., 2005). While supergiant stars usually have $\log(g) \approx 3.0$ (Cox, 2000), our choice was governed by the availability of model spectra. Using spectra from different epochs (Figure 5.1), we infer $T = 31400 \pm 1200$ K and $A_V = 6.5 \pm 0.3$ mag, consistent with Pellizza et al. (2006) values.

We use the model atmospheres with these parameters in the `getvel` software to measure radial velocities (Table 5.1). Owing to the short orbital period and the supergiant nature of the primary, we assume that the orbit is circular. We measure $K_{\rm opt} = 25.0 \pm 1.6$ km s^{-1}, $\gamma_{\rm opt} = -17.6 \pm 1.3$ km s^{-1} and $T_0 = 2553735.77 \pm 0.07$ (HJD), with $\chi^2 = 1.8$ for two degrees of freedom (Figure 5.2). $K_{\rm opt}$ is robust to the choice of model atmospheres: for example, varying T from 27000 K to 33000 K changes $K_{\rm opt}$ by less than 0.5-σ.

Pellizza et al. (2006) infer that the mass of the primary is in the range of $25-28$ M$_\odot$. Plugging this value of $K_{\rm opt}$ into Equation 4.4, we calculate lower limits on $M_{\rm X}$ as a function of $M_{\rm opt}$ (Figure 5.3a). For $M_{\rm opt}$ in the range $25-28$ M$_\odot$ (Pellizza et al., 2006), the lower limits on the NS mass are $M_{\rm X} > 1.79 \pm 0.12$ M$_\odot$ and $M_{\rm X} > 1.93 \pm 0.12$ M$_\odot$ at the extreme values of $M_{\rm opt}$.

We further revise this estimate based on the non-eclipsing nature of this SFXT. For a circular orbit with semi-major axis a and primary stellar radius R, eclipses are seen if $\cos i < R/a$. The semi-major axis is a function of the binary period P and masses of the components. For the O star in IGR J17544-2619, 12.7 R$_\odot < R < 23$ R$_\odot$ (Rahoui et al., 2008; Clark et al., 2009). The lower limit on R can be used to calculate a corresponding upper limit on $\sin i$. Substituting this upper limit in Equations 4.2 or 4.4, we calculate a revised lower limit on $M_{\rm X}$ (Figure 5.3b). For the most conservative scenario ($M_{\rm opt} = 25$ M$_\odot$, $R = 12.7$ R$_\odot$) we calculate $M_{\rm X} > 1.91 \pm 0.13$ M$_\odot$. Using other values in the ($R, M_{\rm opt}$) parameter space results in even stronger lower limits (higher values of $M_{\rm X,min}$).

The existence of such massive neutron stars puts strong constraints on the equation of state of matter.

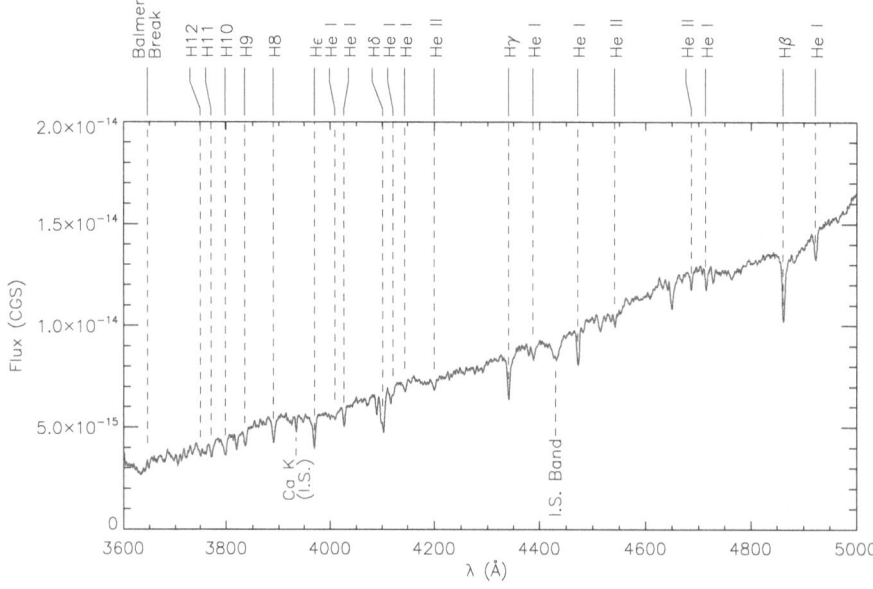

Figure 5.1. Blue side spectrum of the primary in IGR J17544-2619. Major H and He lines are marked. The primary is O9 Ia star, with $T = 31000$ K, and has foreground extinction $A_V = 6.3 \pm 0.4$ (Pellizza et al., 2006).

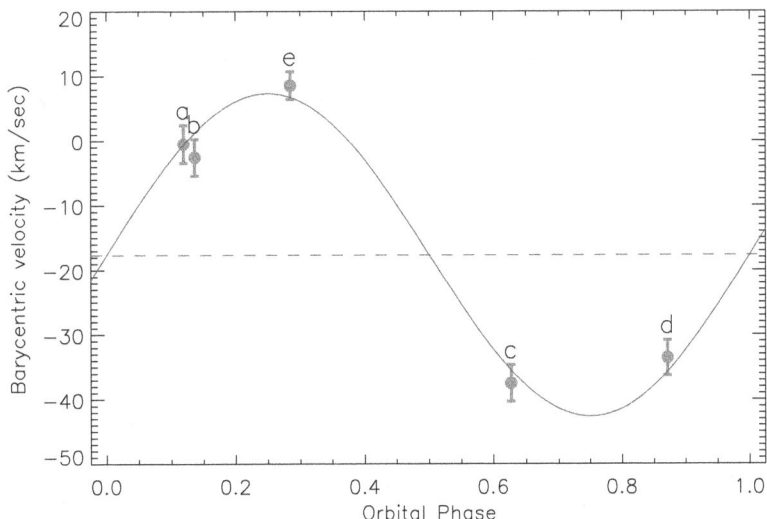

Figure 5.2. Radial velocity fit for IGR J17544-2619. We assume a circular orbit and measure $K_{opt} = 25.0\pm1.6\,\mathrm{km\,s^{-1}}$, $\gamma_{opt} = -17.6 \pm 1.3\,\mathrm{km\,s^{-1}}$ and $T_0 = 2553735.77 \pm 0.07$ (HJD), with $\chi^2 = 1.8$ for two degrees of freedom.

If X-ray pulsations are detected in IGR J17544-2619 in more epochs, we can measure $a_X \sin i$, and in turn the mass ratio q (Equation 4.3). This will provide an additional constraint on M_X, independent of R. Given the significant implications of verifying a high NS mass, X-ray timing observations are of utmost importance. We plan to apply for observing time with *XMM* and NuSTAR to monitor IGR J17544-2619 for pulsations.

5.2   SAX J2103.5+4545

Discovered with *BeppoSAX* in an outburst in 1997, SAX J2103.5+4545 is an an X-ray pulsar with a 358.61 s pulse period (Hulleman et al., 1998). The system is a Be HMXB, consisting of a NS and a B0 Ve star in a 12.7 d orbit. Based on the foreground extinction ($A_V = 4.2 \pm 0.3$) and spectral type, system is at 6.5 ± 0.9 kpc (Reig et al., 2004). We obtained spectra SAX J2103.5+4545 on all six "X-Mas at Palomar" epochs (Figure 5.4). Typical B0 stars have $T = 30000$ K and $\log(g) = 4.0$ (Cox, 2000). Fitting the spectra with (Munari et al., 2005) $\log(g) = 4.0$ model atmospheres as discussed in Section 5.1, we calculate $T = 29200 \pm 700$ and $A_V = 4.0 \pm 0.3$. Using spectra with $T = 30000$ K and $\log(g) = 4.0$ and varying the rotational velocity, we calculate $v_{rot} \sin i = 246 \pm 8$ km s^{-1}, consistent with the past measurement $(240 \pm 20$ km s^{-1}; Reig et al., 2004).

We use model stellar spectra with T=29,000 K, $\log(g)= 4.0$, and $v_{rot} = 250$ km s^{-1} to calculate radial velocities (Table 5.1). Two of the four spectra acquired on 2009 Aug 23 were acquired through variable cloud cover and have low SNR. We excluded these two spectra while calculating the RV for this epoch. On folding the data at the orbital period, the data point "c" from 2009 July 31 is seen to be an outlier (Figure 5.5). Including this point in the fitting procedure does not give any satisfactory solution. We

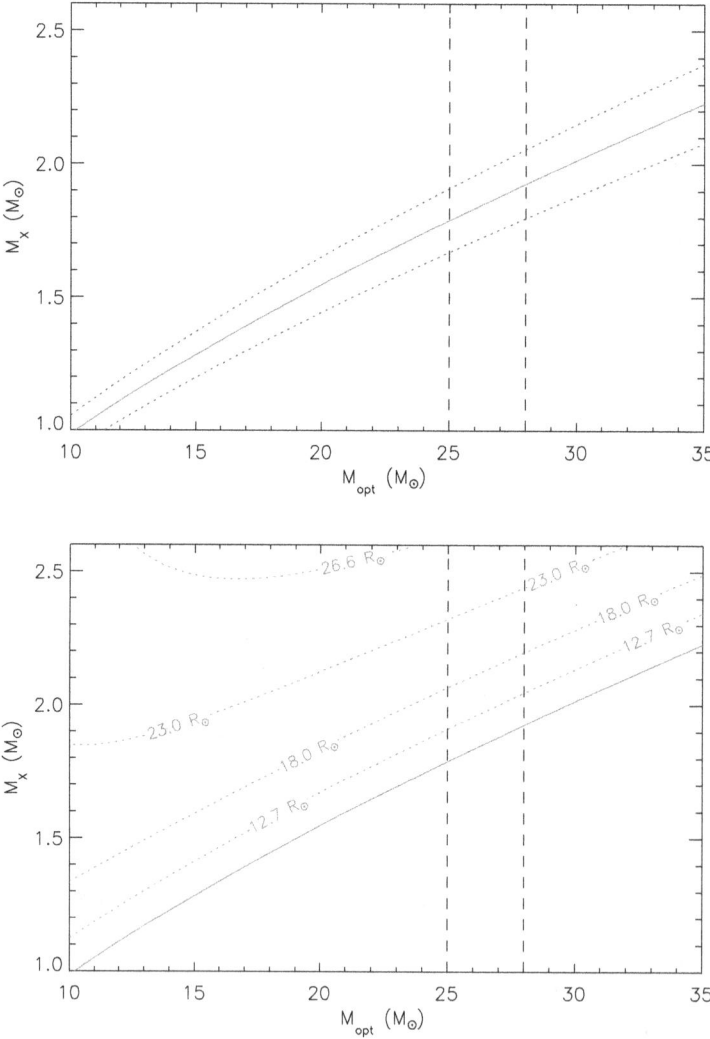

Figure 5.3. Lower limits on M_X for IGR J17544-2619. Top: The solid red curve relates M_X and M_{opt} assuming $\sin i = 1$, for $K_{opt} = 25$ km s^{-1}. The blue curves are ± 1-σ limits corresponding to a velocity uncertainty of $\Delta K_{opt} = 1.6$ km s^{-1}. The vertical black dashed lines at $M_{opt} = 25, 28$ M$_\odot$ are the bounds of M_{opt} from Pellizza et al. (2006). The corresponding lower limits are $M_X > 1.79 \pm 0.11$ M$_\odot$ and $M_X > 1.93 \pm 0.12$ M$_\odot$ respectively. Bottom: The non-detection of eclipses places upper limits on $\sin i$ as a function of the stellar radius R, changing the $M_X - M_{opt}$ relation. The solid line corresponds to $\sin i = 1$ ($R = 0$), and dashed lines are contours of $M_{X,min}$ for various values of R. Adopting $R_{min} = 12.7$ R$_\odot$ and $M_{opt,min} = 25$ M$_\odot$, we deduce $M_X > 1.91 \pm 0.13$ M$_\odot$.

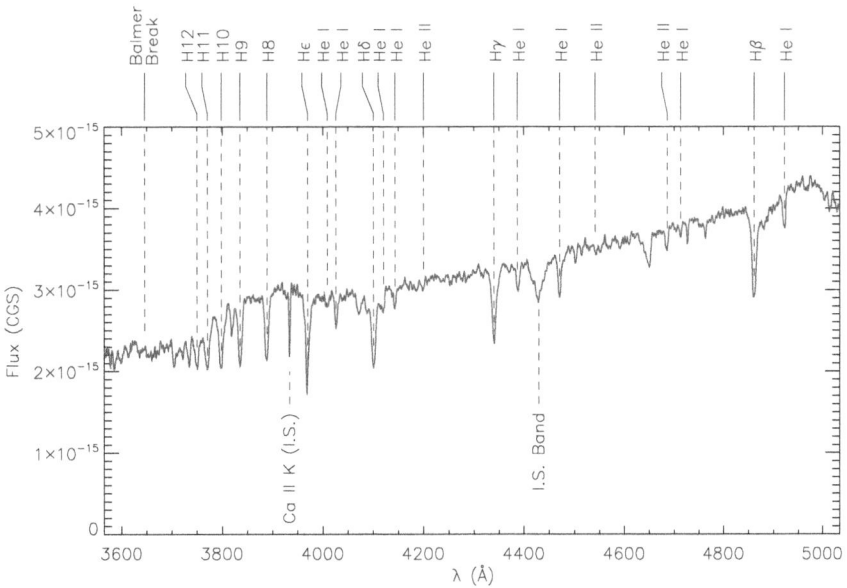

Figure 5.4. Blue side spectrum of the primary in SAX J2103.5+4545. Major H and He lines are marked. The primary is B0 Ve star, with $T = 29200 \pm 700$ K and projected rotational velocity $v_{\rm rot} \sin i = 246 \pm 8$ km s^{-1}. The broad bump in continuum around 3900 Å is an artifact of flatfielding.

examined the observing logs and spectra to look for possible causes of a change in radial velocity, but did not find any. To verify the accuracy of our wavelength solution, we measured the centroid of the interstellar Ca K line, which should be independent of the radial velocity of the target. We found that the location of the line on this epoch is the consistent with other epochs within the ~ 10 km s^{-1} measurement error. We acknowledge this discrepancy but exclude this point while fitting the orbit.

We fit the orbit using the following NS parameters from Camero Arranz et al. (2007):

MJD at periastron $T_0 = 52548.577$
Orbital period $P = 12.66528 \pm 0.00051$
Longitude of periastron $\omega_{\rm X} = 241° \pm 2°$
Eccentricity $e = 0.40 \pm 0.02$

For the B star, $\omega_{\rm opt} = \omega_{\rm X} - 180°$. Using the five data points (a, b, d, e, f), we calculate $\gamma = -72.3 \pm 1.3$ km s^{-1} and $k_{\rm opt} = -12.5 \pm 2.0$ km s^{-1}, with $\chi^2 = 2.7$ for 3 degrees of freedom (Figure 5.5).

Camero Arranz et al. (2007) analyze the times of arrival of X-ray pulses and calculate $a_{\rm X} \sin i = 80.81 \pm 0.67$ lt-s $= (2.42 \pm 0.02) \times 10^7$ km. Following Hilditch (2001, Equations 2.46, 2.50), we convert this to $K_{\rm X}$:

$$K_{\rm X} = \frac{2\pi}{(1 - e^2)^{1/2}} \frac{a_{\rm X} \sin i}{P} \tag{5.1}$$

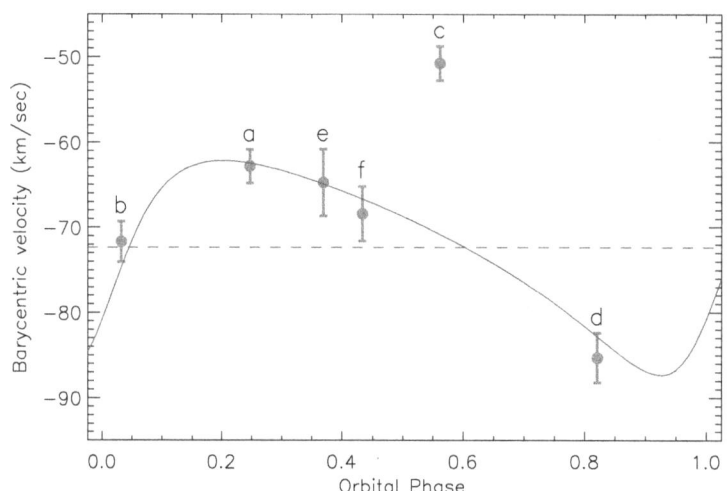

Figure 5.5. Radial velocity fit for SAX J2103.5+4545. I use a model atmosphere with T=29,000 K, $\log(g)$= 4.0, $v_{\rm rot} = 250$ km s^{-1}. T_0 and phase 0 is defined as the periastron passage, corresponding to MJD 52548.577. Orbital period $P = 12.66528 \pm 0.00051$, $\omega_{\rm X} = 241° \pm 2°$, and $e = 0.40 \pm 0.02$ are held fixed at the best fit values by Camero Arranz et al. (2007). Note that $\omega_{\rm opt} = \omega_{\rm X} - 180°$. We calculate $\gamma = -72.3 \pm 1.3$ km s^{-1} and $k_{\rm opt} = -12.5 \pm 2.0$ km s^{-1}. The data point c corresponds is from 2009 July 31 and is excluded from the fit (see text).

If we measure P in days, $a_{\rm X} \sin i$ in km and velocity in km s^{-1}, we get:

$$K_{\rm X} \quad = \quad \frac{7.27 \times 10^{-5}}{(1 - e^2)^{1/2}} \frac{a_{\rm X} \sin i}{P} \tag{5.2}$$

$$= \quad 152 \pm 6 \text{ km s}^{-1} \tag{5.3}$$

For SAX J2103.5+4545, the uncertainty is dominated by the uncertainty in e. Using our best-fit value of $K_{\rm opt}$, we get $q = M_{\rm X}/M_{\rm opt} = K_{\rm opt}/K_{\rm X} = 0.082 \pm 0.013$. Assuming $M_{\rm opt} \approx 17.5$ M$_\odot$, we get $M_{\rm X} = 1.44 \pm 0.14$. Assuming a reasonable range of B0 V star masses ($M_{\rm opt} = 17.5 \pm 2$ M$_\odot$), we conclude $M_{\rm X} = 1.4 \pm 0.3$ M$_\odot$.

5.3   1H 2138+579

1H 2138+579 is a Be HMXB with a 66 s period transient X-ray pulsar. Observations by Bonnet-Bidaud & Mouchet (1998) revealed a B1.5 Ve primary with foreground extinction $A_V = 4.0 \pm 0.3$, corresponding to a distance $d = 3.8 \pm 0.6$ kpc. The measured projected rotational velocity $v_{\rm rot} \sin i = 460$ km s^{-1} is close to the breakup velocities of B1-B2V stars (Bonnet-Bidaud & Mouchet, 1998, and references therein). Assuming that the B star rotation is aligned with the orbit, they conclude $i > 65°$.

The blue side DBSP spectrum of 1H 2138+579 shows Hβ and Hγ in emission (Figure 5.6). The emission components likely arise from a "shell" or the disc around the Be star. Their strength varies from epoch to epoch, and the emission often fills up the broader absorption component from the stellar photosphere. He lines are relatively weak, as expected in a B1–B2 V star.

Typical B1–B2 V stars have $T = 20000-26000$ K, $\log(g) = 4.0$ (Cox, 2000). For velocity measurements, we use Munari et al. (2005) model atmospheres with $T = 24000$ K, $\log(g) = 4.0$, $v_{\rm rot} = 500$ km s^{-1} and solar metallicity. We calculate the foreground extinction to be $A_V = 3.7 \pm 0.2$, and hold it fixed at $A_V = 3.7$ for RV measurements. The measured velocities do not show any clear orbital trend (Table 5.1, Figure 5.7).

Assuming $M_{\rm opt} = 11$ M$_\odot$, $M_{\rm X} = 1.5$ M$_\odot$, $e = 0$ and $\sin i = 0.95$ ($i = 72°$), we expect $K_{\rm opt} \approx 20$ km s^{-1}, comparable to values we have measured for other objects. We attribute the non-detection of this signal to the variability of the H lines and the relatively low strength of other lines. Further analysis is under way to try and circumvent this limitation.

5.4   GRO J2058+42

GRO J2058+42 was discovered by the *Compton Gamma Ray Observatory* as a 198 s pulsar (Wilson et al., 1998). It is a Be HMXB with a 55.03 d period, at distance of 9.0 ± 1.3 kpc. The primary is a subgiant or a dwarf star of spectral class O9.5–B0 (Wilson et al., 2005), with projected rotation velocity $v_{\rm rot} \sin i = 240 \pm 50$ km s^{-1} (Kzlolu et al., 2007). Assuming that the spin of the primary is aligned with the orbit, they constrain the inclination i to be $> 40°$ based on the break–up rotation velocity.

We obtained 36 spectra of GRO J2058+42 over the six epochs (Figure 5.8). We fit Munari et al. (2005) model atmospheres to our observed spectra and calculate $A_V = 4.4 \pm 0.4$, consistent with $A_V = 4.3 \pm 0.3$ measured by Kzlolu et al. (2007). The data are best fit by model atmospheres with $T = 33000 - 35000$ K and $\log(g) = 4.0$, as expected from the spectral type (Cox, 2000). Hence, we use model spectra with $T = 33000$ K, $\log(g) = 4.0$, $v_{\rm rot} = 250$ km s^{-1} and solar metallicity for measuring radial velocities (Table 5.1).

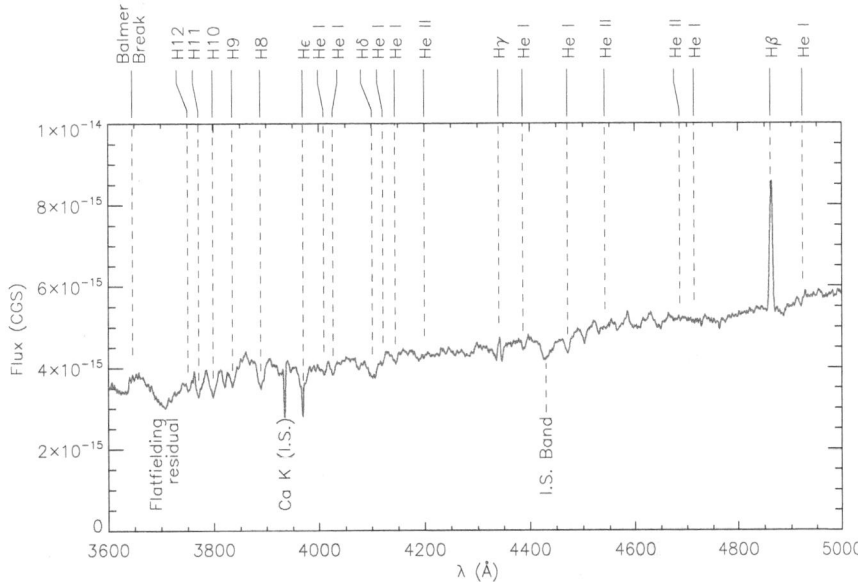

Figure 5.6. Blue side spectrum of the primary in 1H 2138+579. Major H and He lines are marked. Hβ is seen in emission, while Hγ has an emission core.

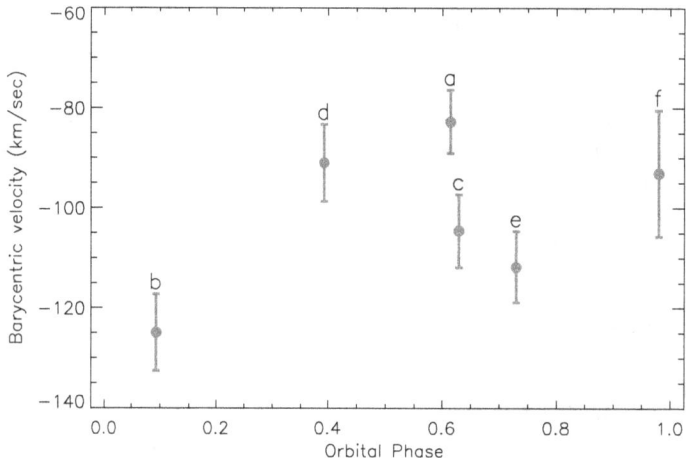

Figure 5.7. Radial velocity measurements for 1H 2138+579. I use a model atmosphere with T=24,000 K, log(g)= 4.0, $v_{\mathrm{rot}} = 500$ km s^{-1}. The orbital period is $P = 20.85$ d. For this plot, data were folded with an arbitrarily selected heliocentric Julian date, $T_0 = 2455015.866$.

On folding the data on the orbital period, it is evident that the orbit is not circular, as is the case with many Be HMXBs. The radial velocity solution to an eccentric orbit has two additional parameters:

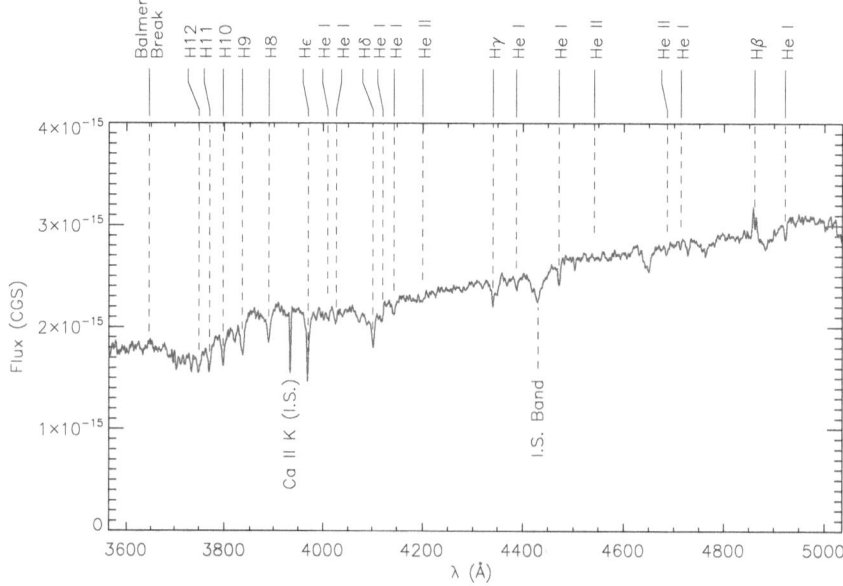

Figure 5.8. Blue side spectrum of the primary in GRO J2058+42. Major H and He lines are marked.

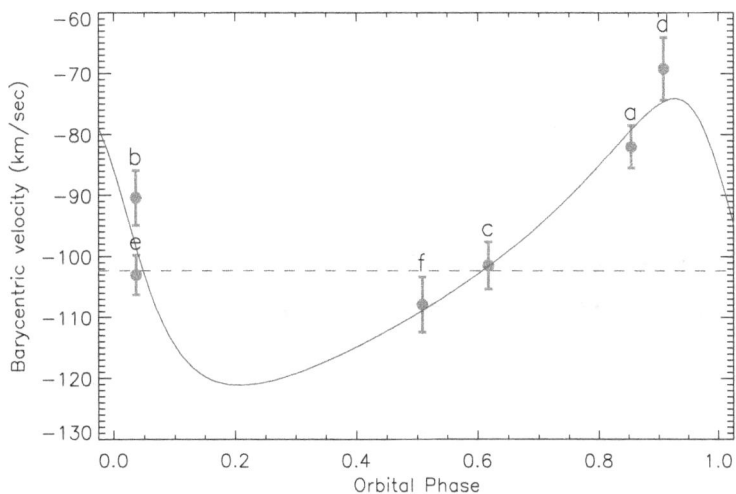

Figure 5.9. Radial velocity measurements for GRO J2058+42. I use a model atmosphere with $T = 33000$ K, $\log(g)$ = 4.0, $v_{\rm rot} = 250$ km s^{-1}. The orbital period is $P = 55.03$ d. For this plot, I assumed $e = 0.4$, $\omega = 60°$, and got $T_0 = 52534.6 \pm 0.5$, $\gamma = -102 \pm 2$, $K_{\rm opt} = 26 \pm 4$. While these are reasonable parameters and values, that is too many assumptions for 6 data points.

eccentricity e and argument of periastron ω. We choose not to fit a five parameter ($K_{\rm opt}$, γ, T_0, e, ω) solution to 6 data points. To demonstrate the feasibility of such an approach, we show a representative fit in Figure 5.9. We *assume* the values $e = 0.4$ and $\omega = 60°$, similar to SAX J2103.5+4545, and calculate other parameters from the RV data. We find $T_0 = 52534.6 \pm 0.5$, $\gamma = -102 \pm 2$, and $K_{\rm opt} = 24 \pm 4$. While these are reasonable values for the parameters, they are calculated by assuming values of e and ω and as such, should not be taken at face value. The preferred approach would be to obtain X-ray timing data to measure e, ω, T_0 and the projected semi-major axis $a_X \sin i$. Combining this information with our RV measurements, we can constrain the mass of this neutron star.

In conclusion, the ongoing "X-Mas at Palomar" project has yielded constraints on masses of two NSs (SAX J2103.5+4545, IGR J17544-2619), and has set the stage for a third (GRO J2058+42). In the coming months, results for the six targets will be compiled into a publication, increasing the sample size of NS–HMXB mass constraints by 50%.

Chapter 6

Constraints on the Compact Object Mass in the Eclipsing HMXB XMMU J013236.7+303228

Varun B. Bhalerao,[a] Marten H. van Kerkwijk,[b] Fiona A. Harrison[a]

[a]Cahill Center for Astrophysics, California Institute of Technology, Pasadena, CA 91125, USA

[b]Department of Astronomy and Astrophysics, University of Toronto, 50St. George Street, Toronto, ON M5S 3H4, Canada.

Abstract

We present optical spectroscopic measurements of the eclipsing High Mass X-ray Binary XMMU J013236.7 +303228 in M 33. Based on spectra taken at multiple epochs of the 1.73 d binary orbital period we determine physical as well as orbital parameters for the donor star. We find the donor to be a B1.5IV sub-giant with rough uncertainty of 0.5 spectral in spectral class and one luminosity class. The effective temperature is $T = 22,000 - 23,000$ K. From the luminosity, temperature and known distance to M33 we derive a radius of $R = 8.9 \pm 0.5 \, R_\odot$. From the radial velocity measurements, we determine a velocity semi-amplitude of $K_{opt} = 63 \pm 12$ km s^{-1}. Using the physical properties of the B-star determined from the optical spectrum, we estimate the star's mass to be $M_{opt} = 11 \pm 1 \, M_\odot$. Based on the X-ray spectrum, the compact companion is likely a neutron star, although no pulsations have yet been detected. Using the spectroscopically derived B-star mass we find the neutron star companion mass to be $M_X = 2.0 \pm 0.4 \, M_\odot$, consistent with the neutron star mass in the HMXB Vela X-1, but heavier than the canonical value of 1.4 $M_\odot$ found for many millisecond pulsars. We attempt to use as an additional constraint that the B star radius inferred from temperature, flux, and distance, should equate the Roche radius, since the system accretes by Roche lobe overflow. This leads to substantially larger masses: $M_X < 2.7^{+0.7}_{-0.6} \, M_\odot$ and $M_{opt} < 18.9^{+2.0}_{-1.9} \, M_\odot$. We find from known systems that by applying this technique in its simplest form the masses are consistently overestimated, and conclude that precise constraints require detailed modeling of the shape of the Roche surface.

A version of this chapter has been submitted to the *Astrophysical Journal*. It is reproduced here with permission from AAS.

6.1 Introduction

The range of possible neutron star masses depends on many factors, such as the initial mass of the progenitor's stellar core, the details of the explosion (in particular mass accretion as the explosion develops), subsequent mass accretion from a binary companion, and the pressure–density relation, or equation of state (EOS), of the neutron star matter. On the low-mass end, producing a neutron star requires the progenitor's core to exceed the Chandrasekhar mass, which depends on the uncertain electron fraction. Theoretical models place this minimum mass in the range $M_{core} \gtrsim 0.9 - 1.3$ M$_\odot$ (Timmes et al., 1996). The largest possible neutron star mass depends on the unknown physics determining the EOS—for example whether kaon condensates or strange matter can from in the interior (See, for example, Lattimer & Prakash, 2005). The highest-mass neutron star to date with an accurate measurement weighs in at 1.97 ± 0.04 $M_\odot$ (Demorest et al., 2010), which already rules out the presence of exotic hadronic matter a the nuclear saturation density (Demorest et al., 2010; Lattimer et al., 2010).

Testing the predictions of supernova models, binary evolution models, and finding objects at the extremes of the mass spectrum require determining neutron star masses in a variety of systems with differing progenitor masses and evolutionary history. Neutron stars accompanying either a high-mass star or another neutron star are thought to have accreted little to no matter over their lifetimes. In contrast, neutron stars in low-mass X-ray binaries and millisecond pulsars, typically in close orbits around a white dwarf, have undergone extended accretion periods that will make the current mass exceed that at birth. Different types of binaries will also have different average neutron star progenitor masses.

High mass X-ray binaries (HMXBs)—binaries containing a neutron star and a massive ($\simeq 20$ M$_\odot$) companion–are particularly interesting systems in which to pursue mass measurements. In most cases the neutron star progenitor will have been more massive than the observed donor star, yielding a relatively high-mass pre-supernova core. Furthermore, the NS mass will be close to the birth mass, since even for Eddington rates $\leq 0.1 M_\odot$ can be accreted in the $\sim 10^7$ yr lifetime of the OB companion. Indeed, among the five HMXB with reasonably secure masses, one (Vela X-1) has $M = 1.8$ M$_\odot$ (Barziv et al., 2001; Quaintrell et al., 2003), indicating that this neutron star may have been born heavy.

Determining NS masses in HMXBs is, however, difficult. In compact object (NS–NS, NS–white dwarf) binaries, highly precise mass measurements can be obtained from relativistic effects like the precession of periastron (Freire et al., 2008) or measurement of the Shapiro delay (Demorest et al., 2010). In HXMBs, however, accurate mass measurements are limited to eclipsing systems where orbital parameters for both the NS and its stellar companion can be measured. For the NS this is done through X-ray or radio pulse timing, and for the companion through radial velocity measurements derived from doppler shifts in the stellar lines. In the event pulsations are not detected, the NS mass can still be determined if good spectra are available to estimate the mass of the optical component. In rare cases where the distance to the binary is known, this provides an independent constraint on the physical scale of the system—for example by calculating the absolute magnitude of the components. However, calculating masses from such constraints is model dependent.

In this paper we present optical spectroscopic measurements of the donor star in the eclipsing HMXB XMMU J013236.7+303228 using the Low Resolution Imaging Spectrograph on the 10 m Keck-I telescope (LRIS; Oke et al., 1995) aimed at determining the mass of the compact companion. XMMU J013236.7+303228 was discovered by Pietsch et al. (2004) in their *XMM-Newton* survey of M 33. In follow-up observations, Pietsch et al. (2006) identified it as an eclipsing High Mass X-ray Binary with a 1.73 d period. The X-ray spectrum is hard, and the shape implies that the compact object is a neutron star. However we did not detect any pulsations in the X-ray data, so a black hole cannot be ruled out.

Shporer et al. (2006) discovered an optical counterpart (Figure 6.1) which shows variability consistent with ellipsoidal modulation of the OB star. Given high quality spectra we are able to obtain a spectroscopic mass for the donor, and therefore determine the compact object mass. Using the known distance to M33 combined with the B stars luminosity and temperature we derive a physical radius, which we equate with the Roche radius based on the observation that accretion is occurring via Roche lobe overflow. This provides an additional orbital constraint that we use to independently estimate the compact object mass.

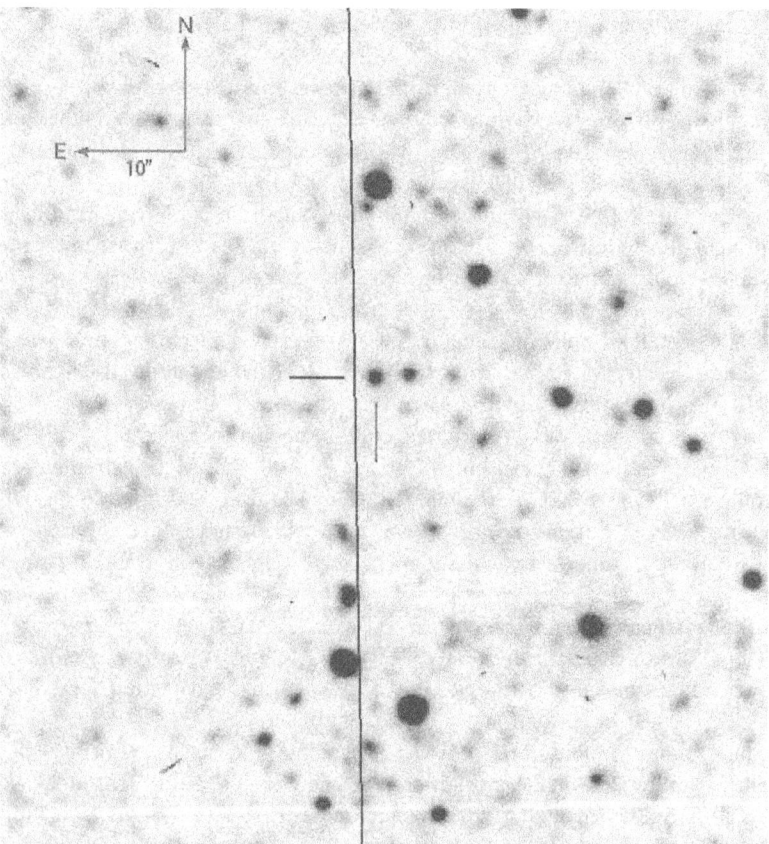

Figure 6.1. A V-band CFHT image showing the optical counterpart to XMMU J013236.7+303228, located at $\alpha = 01^h32^m36.^s94$, $\delta = +30°32'28.''4$ (J2000). The image was obtained from CFHT online data archive, and a WCS was added using astrometry.net (Lang et al., 2010).

6.2   Observations And Data Reduction

We observed XMMU J013236.7+303228 on UT 2009 October 16 and October 17, with the Low Resolution Imaging Spectrograph on the 10 m Keck-I telescope (LRIS; Oke et al., 1995), with upgraded blue (McCarthy et al., 1998; Steidel et al., 2004) and red cameras (Rockosi et al., 2010), covering a wavelength range from 3,200 Å to 9,200 Å. We set up LRIS with the 600/4000 grism on the blue side and the

Table 6.1. Details of individual exposures

Image Name	MJD	Seeing[a] arcsec	SNR	Heliocentric Velocity[b] km s^{-1}	χ^2	Notes[c]
b091016_0062	55120.276	1.0	5.2	-106 ± 24	0.96	Rejected: Logs, Poor SNR
b091016_0063	55120.283	1.0	21.7	-49 ± 5	1.05	Selected
b091016_0064	55120.305	1.0	23.0	-32 ± 5	1.08	Selected
b091016_0066	55120.333	1.0	22.8	-57 ± 5	1.20	Selected
b091016_0080	55120.394	1.0	22.7	-101 ± 6	1.36	Rejected: Logs
b091016_0081	55120.418	1.0	8.7	-109 ± 15	0.99	Rejected: Poor SNR
b091016_0082	55120.425	1.1	20.1	-50 ± 6	1.16	Selected
b091016_0087	55120.457	1.1	13.7	-47 ± 9	1.09	Selected
b091016_0088	55120.474	1.4	4.2	-65 ± 30	1.14	Rejected: Logs, Poor SNR
b091016_0098	55120.501	1.2	18.3	-103 ± 6	1.08	Rejected: Logs
b091016_0099	55120.524	1.1	19.0	-68 ± 6	1.02	Selected
b091016_0106	55120.560	1.2	15.5	-131 ± 8	1.22	Selected
b091016_0108	55120.586	1.1	11.4	-119 ± 11	1.00	Selected
b091016_0109	55120.597	1.1	7.6	-162 ± 19	1.09	Rejected: Poor SNR
b091016_0110	55120.608	1.2	6.3	-128 ± 24	1.03	Rejected: Logs, Poor SNR
b091017_0057	55121.279	0.9	11.3	-42 ± 11	0.99	Rejected: Logs
b091017_0059	55121.305	0.7	18.8	-96 ± 6	0.94	Selected
b091017_0060	55121.327	0.9	19.6	-79 ± 6	0.93	Selected
b091017_0068	55121.367	0.8	22.5	-75 ± 5	0.93	Selected
b091017_0069	55121.389	0.8	22.5	-61 ± 5	0.97	Selected
b091017_0071	55121.419	1.0	19.5	-77 ± 6	0.99	Selected
b091017_0078	55121.457	1.0	19.7	-57 ± 6	0.95	Selected
b091017_0079	55121.478	1.1	16.9	-27 ± 7	1.03	Selected
b091017_0081	55121.507	1.1	18.0	-46 ± 7	0.98	Selected
b091017_0087	55121.541	1.1	17.2	-44 ± 7	1.02	Selected
b091017_0089	55121.567	1.1	15.1	-53 ± 8	1.03	Selected
b091017_0090	55121.589	1.1	8.7	-70 ± 13	0.89	Rejected: Poor SNR
b091017_0091	55121.600	1.1	8.4	-32 ± 15	0.88	Rejected: Poor SNR

Note. — LRIS was set up with a 1″ slit, the D560 dichroic and clear filters on both red and blue arms. To maximize stability, we used the stationary rotator mode. The 600/4000 grism gives a dispersion of 0.61 Å/pix on the blue side. On the red side, we configured the 600/7500 grating at 27°.70 (central wavelength 7151 Å), and get dispersion of 0.80 Å/pix.

[a]Seeing was measured as the median FWHM of Gaussians fitted to the trace of the blue side spectrum at several points.

[b]Errors quoted here do not include an additional 15 km s^{-1} error to be added in quadrature, due to motion of star on the slit.

[c]Spectra were not included in the final analysis if either the observing logs mentioned that the star had moved from the slit, or if the signal-to-noise ratio (SNR) per pixel in the extracted spectrum was below 10.

600/7500 grating on the red side, to get dispersions of 0.6 Å/pix and 0.8 Å/pix respectively Table 6.1. To maximize stability of the spectra, we used the "stationary rotator mode," where the instrument rotator was held fixed near zero degrees rather than tracking the parallactic angle while observing. Atmospheric dispersion was compensated for by the ADC (Atmospheric Dispersion Corrector). We acquired a total of 28 spectra of the target, with exposure times ranging from 300 to 1800 seconds. The spectrophotometric standard EG 247 was observed for flux calibration.

The data were reduced in IRAF[1]. The spectra were trimmed and bias subtracted using overscan re-

---

[1]http://iraf.noao.edu/ .

gions. No flatfielding was applied. Cosmic rays were rejected using L.A.Cosmic (van Dokkum, 2001). Atmospheric lines are stable to tens of meters per second (Figueira et al., 2010), so the wavelength solution for the red side was derived using sky lines for each image. For the blue side, the wavelength solution was derived from arcs taken at the start of the night. The spectra were then rectified and transformed to make the sky lines perpendicular to the trace, to ensure proper sky subtraction. The wavelength solutions for arcs taken at various points during the night are consistent with each other to a tenth of a pixel, with only an offset between different arcs. We corrected for this offset after extracting the spectra, by using the 5577.34 Å [O I] line. The spectra were extracted with APALL, and fluxed with data for EG 247 and the standard IRAF lookup tables. We further tweaked the fluxing by using a EG 247 model spectrum from the HST Calibration Database Archive.[2] We used just one standard spectrum per night, and enabled airmass correction in IRAF during flux calibration.

The final spectrum is shown in Figures 6.2 & 6.3. The signal-to-noise ratio per pixel is >10 for most blue side spectra (Table 6.1). The first night, we experienced some tracking issues with the telescope, so the target did not remain well centered on the slit at all times. A similar problem was experienced for the first exposure on the next night, where the object was at a high airmass. In all following discussions, we reject some such spectra based on observing logs, and spectra with signal-to-noise ratio per pixel of <10 (Table 6.1).

6.3 Donor Star Parameters and Orbit

We determine the best-fit stellar parameters and orbital solution using an iterative technique. First, we estimate a spectral type for the primary (donor) from individual spectra. We use appropriate spectral templates to calculate the orbital solution (Section 6.3.2). Next, we shift spectra to the rest frame and combine them to get a higher quality spectrum. We calculate stellar parameters from this combined spectrum, and use a template spectrum with these refined parameters to recalculate the velocities. In this section, we describe the final iterations of both these steps.

6.3.1 Stellar Parameters

Based on photometry of XMMU J013236.7+303228, Pietsch et al. (2009) estimate that the companion is a 10.9 $M_\odot$ object with $T_{\rm eff} = 33000$ K and $\log(g) = 4.5$, with $\chi^2 = 2.4$ for their best-fit model. They then assume a distance of 795 kpc to M33, and calculate that the star has an absolute magnitude $M_V \sim -4.1$ and the line of sight extinction is $A_V = 0.6$, so derive a stellar radius of 8.0 $R_\odot$.

We deduce the spectral type by comparing our spectra to Walborn & Fitzpatrick (1990) and the Gray spectral atlas.[3] The absence of He II lines (Figure 6.2) implies a spectral type later than O, while the relative strengths of the Mg II 4482 Å/He I 4471 Å lines point to a spectral type earlier than B3. The strength of He I lines, and a weak feature at 4420 Å indicate a spectral type around B1, for main sequence stars. A bump blueward of H8 3889 Å is characteristic of spectral type B2. The weakness of C III 4650 Å, and the relative strengths of C III/O II near 4650 Å refine the spectral type to between B1 and B2 for both dwarfs and giants. Finally, the weakness of Si IV Hδ 4101 Å, gives a spectral type of B1.5. To determine the luminosity class, we note that the O II 4415–4417 Å and Si III 4552 Å lines are present, but are weak as compared to He I 4387. We conclude that the donor is a B1.5IV star, with rough uncertainties of 0.5 spectral classes and one luminosity class.

---

[2]ftp://ftp.stsci.edu/cdbs/current_calspec/ .
[3]http://ned.ipac.caltech.edu/level5/Gray/frames.html .

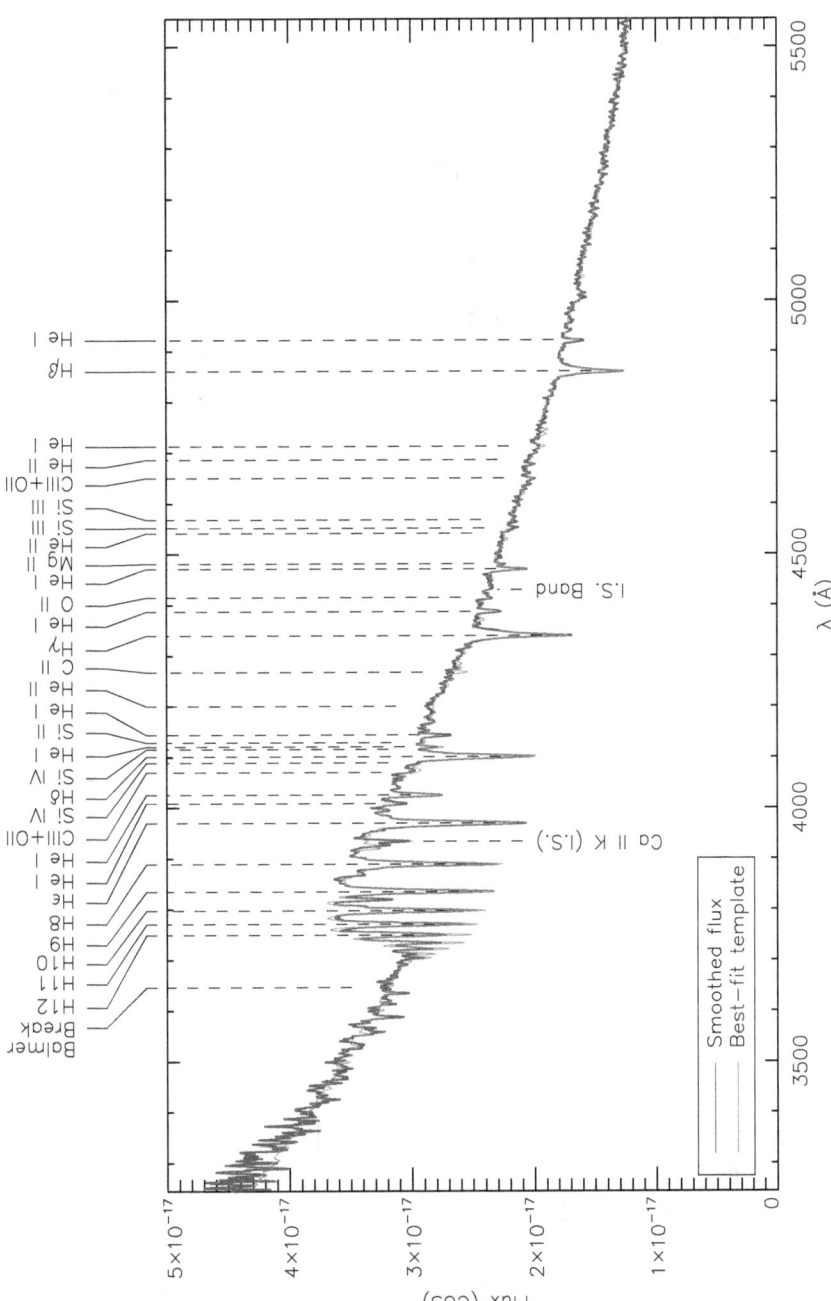

Figure 6.2. Observed spectrum and best-fit model for XMMU J013236.7+303228. The blue line is the average of the ten good spectra obtained on UT 2009 October 17, shifted to the rest wavelength using velocities from Table 6.1. The spectrum is smoothed with a 5 pixel (3 Å) boxcar for plotting. The red line is the best-fit template spectrum with $T = 22000$ K, $\log(g) = 3.5$, $v_{\rm rot} \sin i = 250$ km s^{-1}, solar metallicity. The template is reddened using $A_V = 0.395$, and scaled appropriately. Shifting the spectra to the rest frame blurs out the Ca II interstellar line and the 4430 Å interstellar band. The observed higher Balmer lines are less strong than those of the model, suggesting that the surface gravity is slightly higher than $\log g = 3.5$ (consistent with our estimated spectral type and with our fit results; see text).

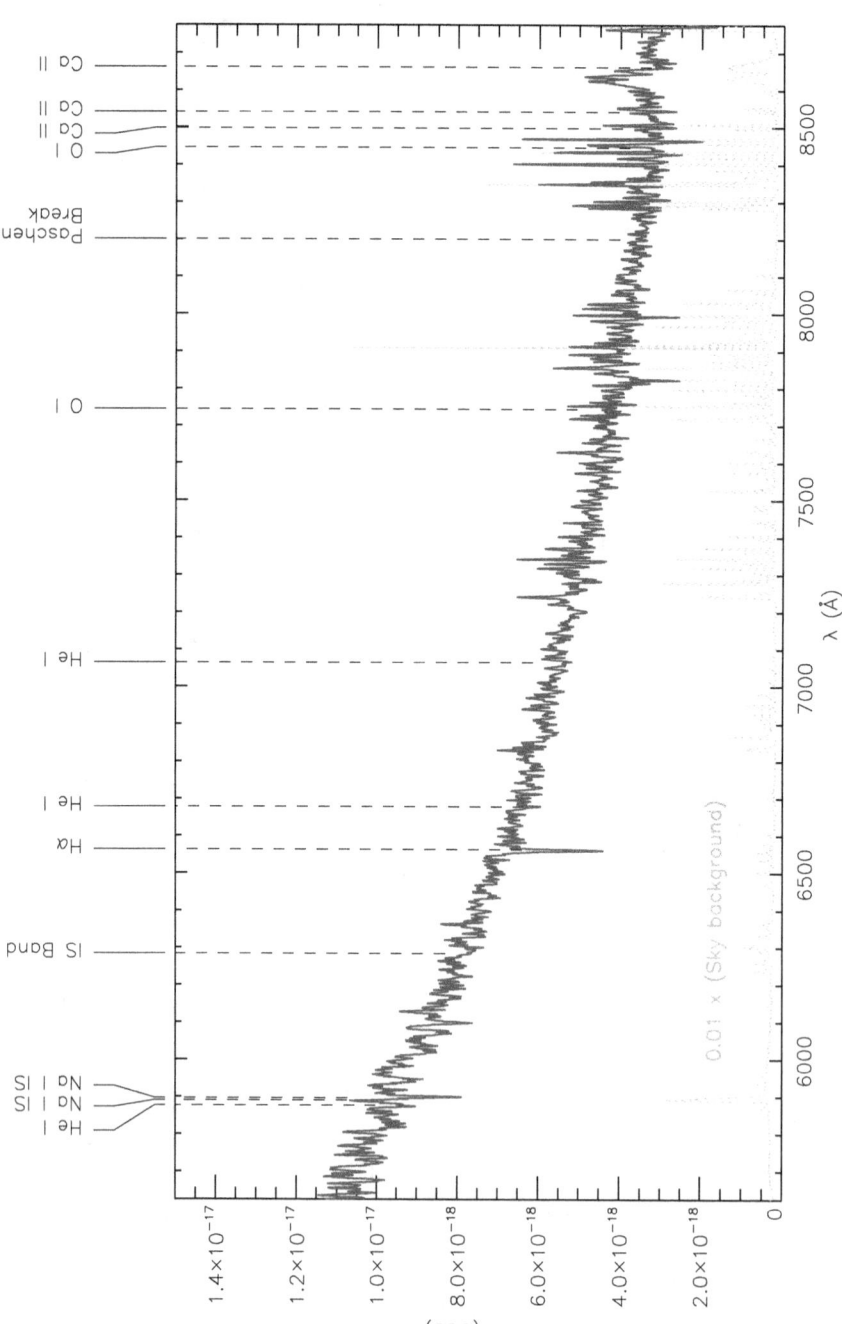

Figure 6.3. Observed red spectrum for XMMU J013236.7+303228. The blue line is the median combination of 48 spectra obtained on UT 2009 October 17, shifted to the rest wavelength using the velocity solution from Table 6.2. The flux axis differs from Figure 6.2. The spectrum is smoothed with a 5 pixel (4 Å) boxcar for plotting. The source is much fainter than the sky in this wavelength region. For reference, the green dashed line shows the extracted sky spectrum, scaled *down* by a factor of 100 for plotting. We do not use the red side spectra for any fitting.

Tabulated values of stellar parameters are usually provided for luminosity class III and V stars. For B1.5V stars, $T \approx 23000$ K, $\log(g) = 4.14$, and $M_V = -2.8$. For B1.5III stars, $T \approx 22000$ K, $\log(g) = 3.63$, and $M_V = -3.4$ (Cox, 2000). The B1.5IV target will have values intermediate to these. This inferred temperature is significantly cooler than $T \sim 33000$ K reported by Pietsch et al. (2009). However, the absence of He II lines at, e.g., 4541 and 4686 Å, clearly exclude such a high temperature.

The colors of a star depend on its temperature and surface gravity. These expected colors can be compared to the observed colors to calculate the reddening and extinction. The expected color is $(B-V)_0 = -0.224$ for a 22000 K sub-giant star, and $(B-V)_0 = -0.231$ for a 23000 K main sequence star (Bessell et al., 1998). From Pietsch et al. (2009), the mean magnitudes are $m_{g'} = 21.03 \pm 0.02$, $m_{r'} = 21.36 \pm 0.02$. Using the Jester et al. (2005) photometric transformations for blue, $U - B < 0$ stars,[4] $m_V = 21.21 \pm 0.03$, $(B-V)_{\rm obs} = -0.09 \pm 0.04$. The color excess is $E(B-V) = (B-V)_{\rm obs} - (B-V)_0 = 0.14 \pm 0.04$. Using the standard ratio of total-to-selective extinction, $R_V = 3.1$ we get $A_V = 0.43 \pm 0.12$. For comparison, the foreground extinction to M33 is $A_V = 0.22$.

We measure various stellar parameters by fitting our combined, fluxed spectrum with model atmospheres (taking into account the instrumental broadening; see Section 6.3.2). For a Roche lobe filling companion (discussed in Section 6.4), we expect a radius of 6–10 $R_\odot$, surface gravity $\log(g) \simeq 3.7$ (consistent with our luminosity class), and projected rotational velocity $v_{\rm rot} \sin i \simeq 250$ km s^{-1}. We use these values as starting points to select model atmospheres from a grid calculated by Munari et al. (2005). These templates are calculated in steps of 0.5 dex in $\log(g)$, so we use models with $\log(g) = 3.5, 4.0$. For the initial fit, we assume $v_{\rm rot} \sin i = 250$ km s^{-1} and solar metallicity. The only free parameters are a normalization and an extinction. We use extinction coefficients from Cox (2000), assuming $R_V = 3.1$. We find that the best-fit model for $\log(g) = 3.5$ has $T = 22100 \pm 40$ K and $A_V = 0.401(3)$, with $\chi^2/{\rm DOF} = 1.13$ for 3600 degrees of freedom, while for $\log(g) = 4.0$, we get $T = 23500 \pm 50$ K and $A_V = 0.425(3)$ with $\chi^2/{\rm DOF} = 1.22$. The temperatures are consistent with those expected for a B1.5 subgiant star, and the extinction is within the range derived from photometric measurements. Pietsch et al. (2009) obtained a higher extinction for the target, which explains why they estimated the source temperature to be higher. Munari et al. (2005) templates are calculated in temperature steps of 1000 K in this range. For further analysis, we use the best-fit template: $\log(g) = 3.5$ and $T = 22000$ K. Since $\log(g)$ is slightly higher than this value, for completeness we also give results using the best fit template for $\log(g) = 4.0$, which has $T = 23000$ K. For both these templates, the best-fit extinction is $A_V = 0.395(3)$. We then keep T and $\log(g)$ constant and vary $v_{\rm rot} \sin i$. For both the $\log(g)$, T combinations, we measure $v_{\rm rot} \sin i = 260 \pm 5$ km s^{-1}. Finally, using the same templates but with varying metallicity, we get the best fits for [M/H] = 0. The 0.5 dex steps in [M/H] are too large to formally fit for uncertainties.

Next, we calculate the luminosity of the object to obtain a radius–temperature relation. The distance modulus to M 33 is $(m - M)_{\rm M33} = 24.54 \pm 0.06$ ($d = 809$ kpc; McConnachie et al., 2005; Freedman et al., 2001), which gives, accounting for the reddening of $A_V = 0.4$, $M_V = -3.74 \pm 0.07$. The bolometric luminosity of a star is related to its temperature and radius by $L_{\rm bol} \propto R^2 T^4$. To obtain the visual luminosity, one must apply a temperature-dependent bolometric correction, $BC = M_{bol} - M_V$. Torres (2010) give formulae for bolometric correction as a power series in $\log(T)$. We calculate $BC = -2.11(-2.21)$ for $T = 22000(23000)$ K (which are consistent with Bessell et al. (1998) tables for main sequence stars).

---

[4] $V = g - 0.59(g - r) - 0.01 \pm 0.01$; $B - V = 0.90(g - r) + 0.21 \pm 0.03$.

After some basic algebra, we obtain:

$$5 \log \left(\frac{R}{R_\odot} \right) + 10 \log \left(\frac{T}{T_\odot} \right) + BC(T) = M_{bol,\odot} - m_V + (m - M)_{M33} + A_V \tag{6.1}$$

$$= 8.48 \pm 0.07.$$

Here, $M_{bol,\odot} = 4.75$ (Bessell et al., 1998). The resultant radius-temperature relationship is shown in Figure 6.4. For $T = 22000(23000)$ K, we infer $R = 9.1(8.7) \pm 0.3 \, R_\odot$. The absolute magnitude and radius are both consistent with a B1.5 sub-giant.

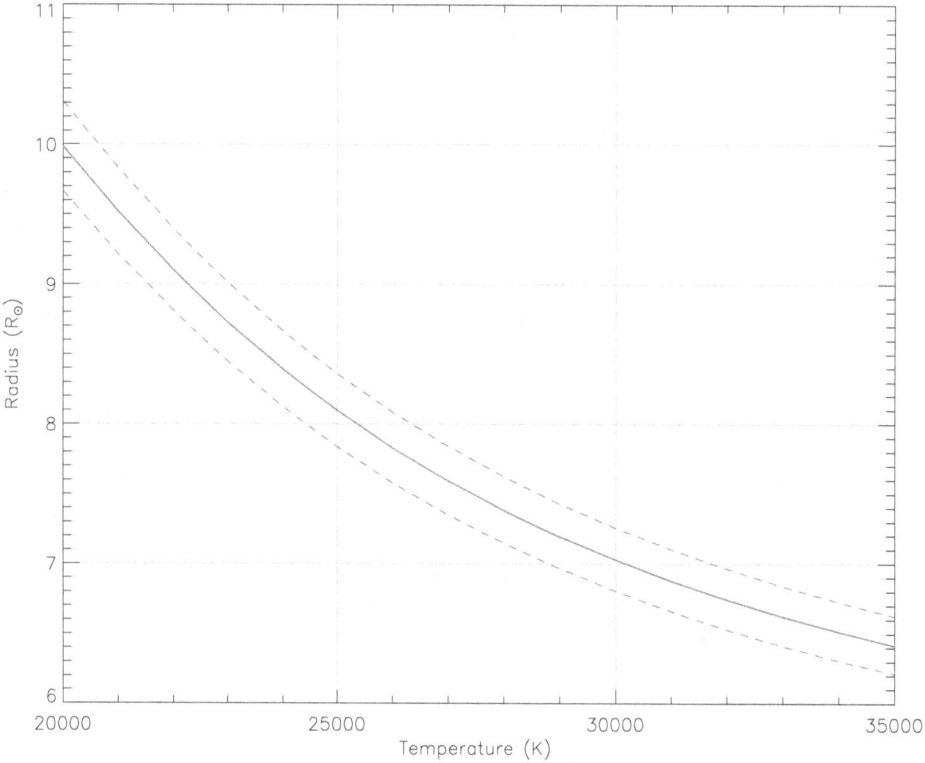

Figure 6.4. Radius–temperature constraint from the observed luminosity. We calculate the absolute visual magnitude from the apparent magnitude, distance modulus and best-fit extinction. We then apply a temperature-dependent bolometric correction to calculate the bolometric magnitude. For any given temperature in this range, the uncertainty in radius is about 3%.

6.3.2 Orbital Parameters

We measure the radial velocities of the B star using model stellar spectra by Munari et al. (2005), as follows. For each observed spectrum, we measure the seeing using the width of the spectral trace. We then generate an instrument response function by taking a Gaussian matched to the seeing, truncating it at the slit size, and convolving it with the pixel size. Fluxed synthetic spectra (templates) are convolved with this instrument response, then redshifted to a test velocity. Then we redden the template using the measured value of extinction, $A_V = 0.395$, with coefficients from Cox (2000). We use IDL[5] mpfit (Markwardt, 2009) to calculate the reddening and normalization to match this spectrum with the observed spectrum, and measure the χ^2. By minimizing the χ^2 over test velocities, we find the best-fit velocity and the error bars. This velocity is converted to a barycentric radial velocity using the baryvel routine in Astrolib (Landsman, 1993). Table 6.1 lists the radial velocities for all spectra, measured using the best-fit stellar template: $T = 22000$ K, $\log g = 3.5$, $[M/H] = 0.0$ and $v_{\rm rot} \sin i = 250$ km s^{-1}.

Red side spectra are not useful for radial velocity measurement for several reasons. The flux of the B star at redder wavelengths is lower than the blue wavelengths, and there are fewer spectral lines in this range. Also, the background noise is higher, from the large number of cosmic rays detected by the LRIS red side and from intrinsic sky emission. Hence, all further discussion omits red side spectra.

We calculate an orbital solution for the B star using these radial velocities. Owing to the short 1.73 d period of the system, we assume that the orbit must be circularized. The orbital solution is then given by:

$$v(t) = \gamma_{\rm opt} + K_{\rm opt} \sin\left(2\pi \frac{t - T_0}{P}\right) \tag{6.2}$$

where $\gamma_{\rm opt}$ is the systemic velocity, $K_{\rm opt}$ is the projected semi-amplitude of radial velocity, and T_0 is the epoch of mid-eclipse. We adopt $T_0 = 2453997.476 \pm 0.006$ from Pietsch et al. (2009). We obtain $\gamma_{\rm opt} = -80 \pm 5$ km s^{-1}, and $K_{\rm opt} = 64 \pm 12$ km s^{-1}. The best-fit has $\chi^2/{\rm DOF} = 5.1$ for 16 degrees of freedom, which is rather poor. We find that an additional error term $\Delta v = 15$ km s^{-1} needs to be added in quadrature to our error estimates to obtain $\chi^2_{\rm red} = 1$. We attribute this to movement of the star on the slit. If the target has systematic offset of $0''.1$ from the slit center over the entire 30 min exposure, it shifts the line centroids by about 0.45 pixels or 18 km s^{-1}. For comparison, van Kerkwijk et al. (2011) find a similar scatter (13 km/s^{-1}) in their observations of a reference star when using LRIS with a similar configuration (600/4000 grating, $0''.7$ slit). Future observations should orient the spectrograph slit to obtain a reference star spectrum to correct for such an offset.

When we fit Equation 6.2 to data including the 15 km s^{-1} error in quadrature, we obtain $\gamma_{\rm opt} = -80 \pm 5$ km s^{-1}, and $K_{\rm opt} = 63 \pm 12$ km s^{-1} (Figure 6.5, Table 6.2). For the epoch of observations, the uncertainty in phase is 0.019 d. If we allow T_0 to vary, we get $T_{0,\rm fit} = 2453997.489 \pm 0.019$ (Heliocentric Julian Date), $\gamma_{\rm opt} = -77.8 \pm 2.1$ km s^{-1}, and $K_{\rm opt} = 59 \pm 5$ km s^{-1}. These values are consistent with those obtained using the Pietsch et al. (2009) ephemeris. Hence, for the rest of this paper, we simply assume their best-fit value for T_0.

To investigate the sensitivity of the result to the choice of the stellar template, we repeat the measurement with a variety of templates. We vary the temperature from O9 (33,000 K) to B3 (18,000 K) spectral classes. As before, we use templates with $\log(g) = 3.5, 4.0$; $v_{\rm rot} \sin i = 250$ km s^{-1} and solar metallicity. Repeating the orbit calculations for each of these models, we find that the systemic velocity $\gamma_{\rm opt}$ may change between models being fit: the extreme values are -77 ± 6 km s^{-1} and -90 ± 5 km s^{-1}, a 1.7-σ difference. We suspect that the reason for this variation is the difference in shape of the continuum, as the extinction was held constant in these fits. For hotter templates with a steeper continuum, the red side of a line has lower flux than the blue side, so lowest χ^2 will be obtained at a slightly higher redshift, as seen. The magnitude of this effect should be independent of the intrinsic Doppler shift of the spectrum, and

---

[5]http://www.ittvis.com/ProductServices/IDL.aspx

Table 6.2.   System parameters for XMMU J013236.7+303228

Property	Value
From Pietsch et al. (2004)	
Period (P)	1.732479 $\pm$0.000027
HJD of mid-eclipse (T_0)	2453997.476 $\pm$0.006
Eclipse half-angle (θ_e)	30°.6 $\pm$1°.2
This work	
Systemic velocity $(\gamma_{\rm opt})$	-80 $\pm$5 km s^{-1}
Velocity semi-amplitude $(k_{\rm opt})$	63 $\pm$12 km s^{-1}
HJD of mid-eclipse (T_0)	2453997.489 $\pm$0.019
Spectroscopically inferred	
OB star spectroscopic mass $(M_{\rm opt})$	11 $\pm$1 M$_\odot$
NS mass $(M_{\rm X})$	2.0 $\pm$0.4 M$_\odot$
Distance-based calculations	
OB star mass $(M_{\rm opt})$	18. $^{+2.0}_{-1.9}$ M$_\odot$
NS mass $(M_{\rm X})$	2.7 $^{+0.7}_{-0.6}$ M$_\odot$

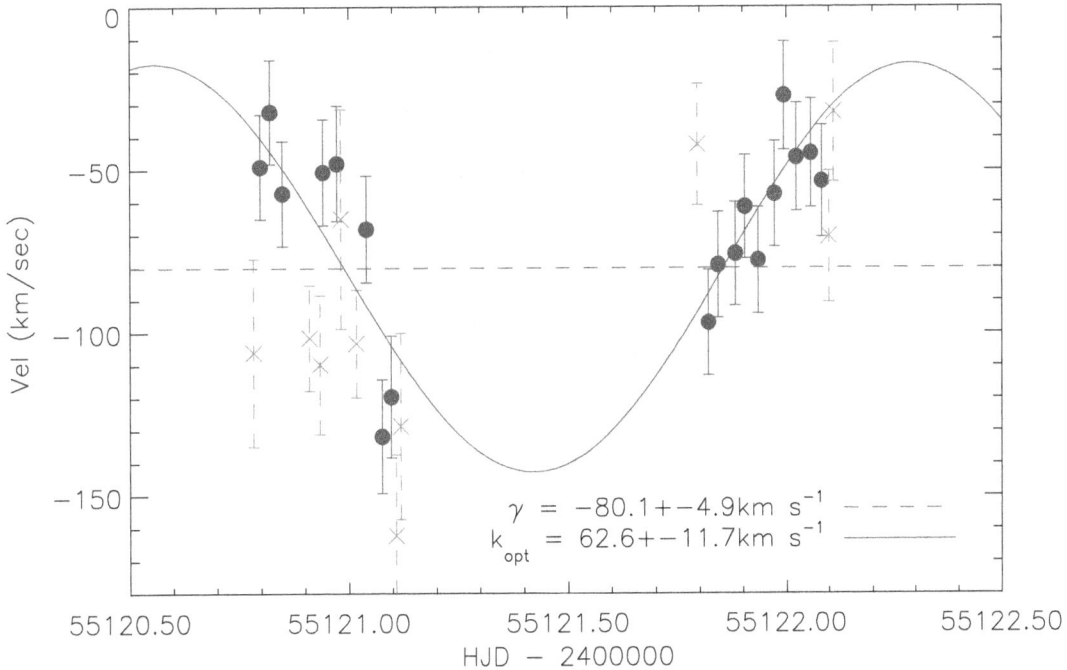

Figure 6.5. The radial velocity curve for XMMU J013236.7+303228. Solid blue circles denote points which were used in the final fit, and gray crosses are points which were rejected based on observing logs or a signal-to-noise ratio under 10 (Table 6.1). The solid blue curve is the best-fit orbital solution, and the dashed blue line is the mean systemic velocity.

should not affect the velocity semi-amplitude K_{opt}. This is indeed the case: K_{opt} is constant irrespective of templates. The extreme values are 62 ± 12 km s^{-1} and 63 ± 12 km s^{-1}, differing by less than 0.1σ. Dynamical calculations depend only on K_{opt}, hence they are robust to the selection of template.

6.4 Component Masses

The general method for accurately determining masses in X-ray binaries (Joss & Rappaport, 1984) requires measuring the orbit for both components, as well has having a constraint on the orbit inclination. In general, the mass (M_1) of a component is expressed in terms of five parameters: the orbital period P, the radial velocity semi-amplitude of the companion (K_2), the eccentricity e, the orbital inclination i and mass ratio $q = M_1/M_2$. The first three parameters can be readily obtained by characterizing the orbit of either component through pulse timing of the NS in the X-ray or radio, or by spectroscopically measuring the radial velocity of the optical companion at optical or infra-red wavelengths. Determining the mass ratio requires measuring orbital parameters for both components: $q = K_2/K_1$. In an eclipsing system, the inclination can be constrained to be nearly edge-on ($i \approx 90°$), with a lower limit derived from eclipse duration and Roche-lobe arguments. Using these measurements, masses of both components in the system can be directly determined without model assumptions (see for example, van der Meer et al., 2007;

Mason et al., 2011b).

Because no pulsations have been detected from the compact object in XMMU J013236.7+303228, we need one more constraint in addition to the radial velocity semi-amplitude of the donor. In Section 6.4.1 we use the spectroscopically inferred mass of the secondary to calculate mass of the secondary from the mass ratio. Because we know the distance to M33, and hence to XMMU J013236.7+303228, we can also estimate the physical size of the secondary from the distance, its luminosity and temperature. This provides a cross check on the mass determination derived from the spectroscopic donor mass (Section 6.4.2).

6.4.1   The Spectroscopic Method

In the following, we will denote the masses of the compact object and donor star by M_X and M_{opt} respectively. The mass of the compact object (M_X) is related to the radial velocity of the B star (K_{opt}) as follows:

$$M_X = \frac{K_{opt}^3 P (1 - e^2)^{3/2}}{2\pi G \sin^3 i} \left(1 + \frac{1}{q}\right)^2, \tag{6.3}$$

where $q = M_X/M_{opt}$ is the ratio of masses, defined so that higher values of M_X relate to higher values of q. P is the orbital period of the binary, e is the eccentricity, and i is the inclination of the orbit. For eclipsing systems, the inclination is constrained by,

$$\sin i = \frac{\sqrt{1 - \beta^2 \left(R_L/a\right)^2}}{\cos \theta_e}, \tag{6.4}$$

where R_L is the volume radius of the Roche lobe, a is the semi-major axis, and β is the Roche lobe filling factor (Joss & Rappaport, 1984). For XMMU J013236.7+303228, the eclipse half-angle is $\theta_e = 30°.6 \pm 1°.2$ (Pietsch et al., 2009). Owing to the short orbital period, we assume that the orbit is circular and the B star rotation is completely synchronized with its orbit. For co-rotating stars, Eggleton (1983) expresses R_L/a in terms of q:

$$\frac{R_L}{a} = \frac{0.49q^{-2/3}}{0.6q^{-2/3} + \ln(1 + q^{-1/3})}. \tag{6.5}$$

The constant, relatively high X-ray luminosity, sustained over the non-eclipsed parts of the orbit, strongly indicates that accretion is occurring via Roche lobe overflow. In Roche lobe overflow, matter flowing through the Lagrangian point may form a disc around the compact object before being accreted onto it. This disc can occult the compact object, causing periods of low X-ray luminosity. Both these characteristics are seen in the X-ray lightcurves of XMMU J013236.7+303228 (Pietsch et al., 2009). If mass is being accreted onto an object in a spherically symmetric manner, the accretion rate is limited by the Eddington rate, $\dot{M}_{Edd}$ and the peak luminosity is $L_{Edd}/L_\odot = 3 \times 10^4 M/M_\odot$. For a $1.5 - 2.5 \, M_\odot$ compact object, $L_{Edd} = 1.8 - 3 \times 10^{38}$ erg s^{-1}. At its brightest, the source luminosity in the $0.2 - 4.5$ keV band was 2.0×10^{37} erg s^{-1}—about $0.1 L_{Edd}$. This luminosity was sustained throughout Chandra ObsID 6387, which covered about 0.72 d of the non-eclipsed orbit (Pietsch et al., 2009). Comparable flux was observed in the non-eclipsed parts of the orbit (0.73 d) in Chandra ObsID 6385. Such high luminosity sustained over significant parts of the orbit is not observed in wind-fed systems, which have typical luminosities an order of magnitude smaller. Further, the short 1.73 d orbital period is not consistent with Be X-ray binary or wind-fed systems. We conclude therefore the B star fills its Roche lobe.

The Roche lobe radius as a function of B star mass is plotted in Figure 6.6 for various neutron star

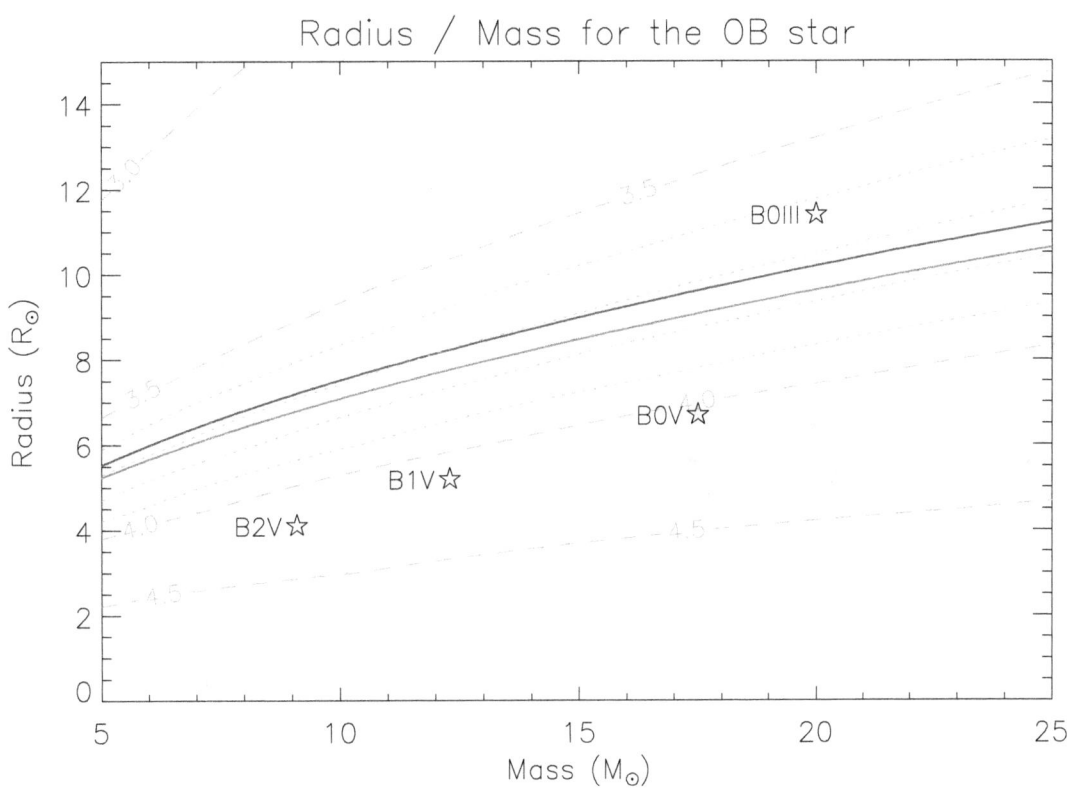

Figure 6.6. Mass–radius relation for the OB star. The solid lines show the Roche lobe radius for the OB star, assuming a 2.4 $M_\odot$ (upper blue line) and a 1.4 $M_\odot$ (lower red line) neutron star. The dashed green lines are contours for $\log(g)$ in steps of 0.5 dex, for which Munari et al. (2005) synthetic spectra are available. In the range $3.5 < \log(g) < 4.0$, dotted $\log(g)$ contours are separated by steps of 0.1 dex. The stars denote the masses and radii of typical isolated B stars.

masses. For a Roche-filling companion, we infer $3.6 \leq \log(g) \leq 3.8$. For a NS mass in the range 1.4–2.4 $M_\odot$ and B star mass 8–20 $M_\odot$, the B star radius lies in the range 6–10 $R_\odot$. Further, the assumed synchronous rotation requires the surface rotational velocity to be in the range 200 km s^{-1} $\lesssim v_{rot} \lesssim$ 285 km s^{-1}. These values are consistent with those derived in Section 6.3.1.

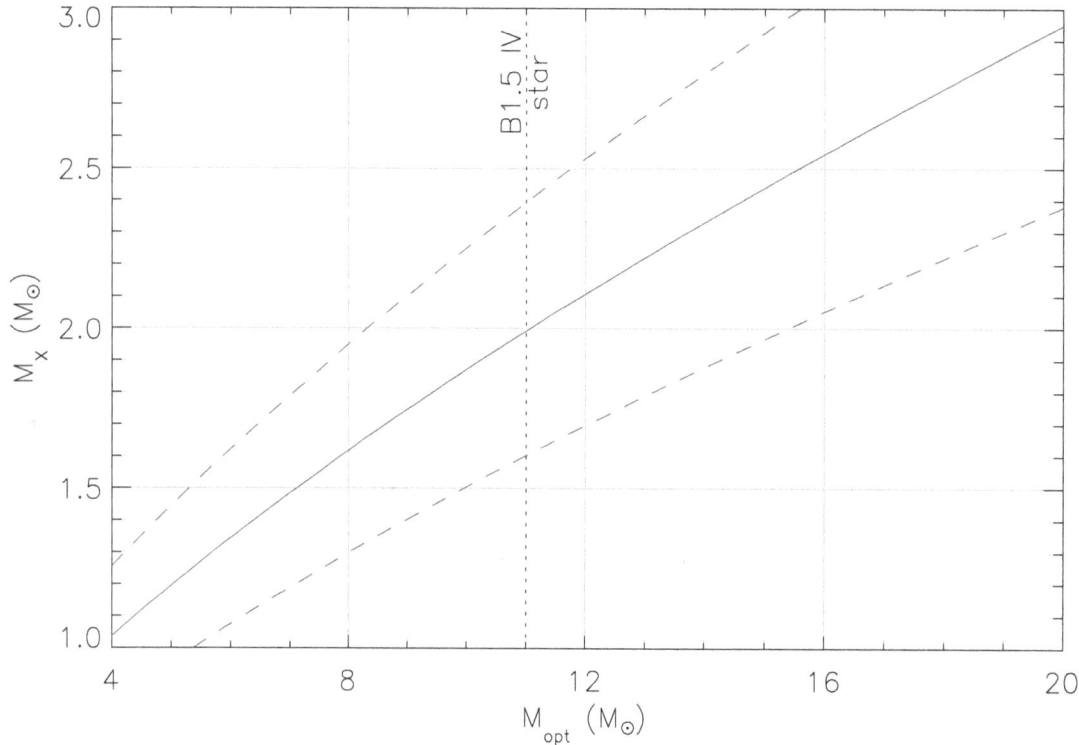

Figure 6.7. The solid blue line shows the compact object mass (M_X) as a function of OB star mass (M_{opt}). Dashed blue lines are ±1-σ errors. The estimated mass of the B1.5IV primary is 11 $M_\odot$, marked by the vertical dotted line. The corresponding mass of the neutron star is 1.86 ± 0.16 $M_\odot$.

Figure 6.7 plots M_X as a function of M_{opt}, calculated by solving Equations (6.3), (6.4) and (6.5) under known constraints. If we know M_{opt}, we can calculate M_X. We use the physical properties of the primary (Section 6.3.1) to estimate the mass of the primary by comparing it with stellar evolutionary models, assuming that binary evolution has not drastically changed the mass-luminosity relation. First, we place the primary on a HR diagram (Figure 6.8) using models by Brott et al. (2011). We conservatively allow for a 0.5 mag error in luminosity. We also plot evolutionary tracks on a $\log(g)$-T figure, to utilize the stricter constraints on $\log(g)$ from Roche lobe arguments. From these plots, we see that the primary is approximately a 15 Myr old, ~ 11 $M_\odot$ star. This is consistent with the typical mass of a B1.5IV star (Cox, 2000).

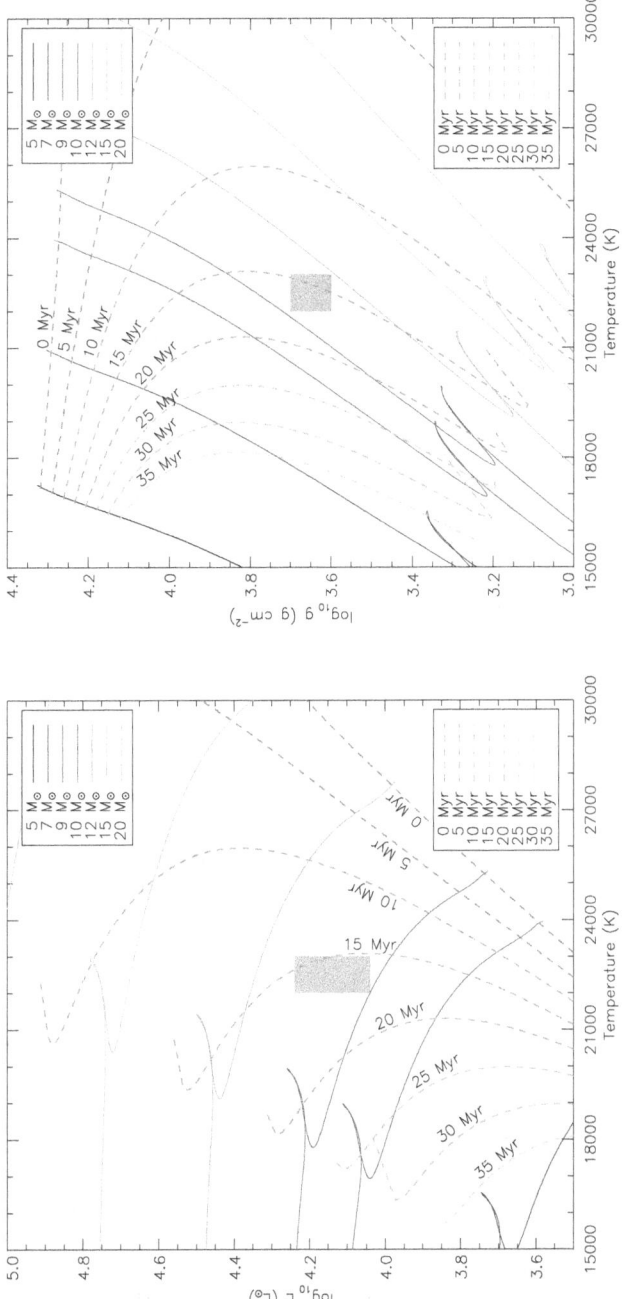

Figure 6.8. Evolutionary tracks and isochrones for massive stars (adapted from Brott et al., 2011). Left panel: the conventional HR diagram with luminosity and temperature. The primary star has 22000 K $\lesssim T \lesssim$ 23000 K. We calculate luminosity from the observed m_V, and allow a 0.5 mag offset to calculate the lower limit (see discussion in Section 6.4.1). This region is shown by a shaded gray box. Right panel: same, but plotted as $\log(g)$ versus T. $\log(g)$ for the primary is constrained from Roche lobe arguments (Section 6.3.1). From both panels, we see that the primary is consistent with a 11 $M_\odot$, 15 Myr object.

For $M_{opt} = 11$ $M_\odot$, we calculate $M_X = 2.0 \pm 0.4$ $M_\odot$. From evolutionary tracks, we estimate that the uncertainty in M_{opt} is ~ 1 $M_\odot$ (Figure 6.8), corresponding to $\Delta M_X = 0.12$, much smaller than the uncertainty arising from ΔK_{opt}. Adding this in quadrature with with uncertainty in the M_X – M_{opt} conversion, we conclude $M_X = 2.0 \pm 0.4$ $M_\odot$.

6.4.2　Masses From Roche Lobe Constraints

In Section 6.3.1, we calculated the radius of the primary from its apparent magnitude, temperature and the distance to M33 (Equation (6.1)). Since the primary is filling its Roche lobe, the stellar radius is equal to the Roche lobe radius (R_L). This additional constraint can be used in Equations (6.3) – (6.5) to solve for M_X and M_{opt}.

We calculate the probability density function (PDF) of component masses as follows. For every pair of assumed masses (M_X, M_{opt}), we use the period P to calculate the semi-major axis a. Then we calculate R_L/a from Equation (6.5) and substitute it into Equation (6.4) to calculate $\sin i$. Using P, a and $\sin i$ we calculate the expected semi-amplitude of the radial velocity:

$$K_2 = \frac{2\pi a \sin i}{P} \frac{M_X}{(M_X + M_{opt})}. \tag{6.6}$$

Next, we calculate the probability for obtaining a certain value of R_L and K_2, given the measured radius R (Section 6.3.1) and K_{opt} (Section 6.3.2):

$$P(R_L, K_2) = \exp\left(-\frac{(R_L - R)^2}{2 \cdot \Delta R^2}\right) \exp\left(-\frac{(K_2 - K_{opt})^2}{2 \cdot \Delta K_{opt}^2}\right) \tag{6.7}$$

Here, we are making a simplifying assumption that the Roche volume radius (Equation (6.5)) is same as the effective radius from photometry (Equation (6.1)). We convert this PDF to a probability density as a function of M_X, M_{opt} by multiplying by the Jacobian $\partial(R, K_{opt})/\partial(M_X, M_{opt})$.

The results are shown in Figure 6.9. Red contours show the 68.3% and 95.4% confidence intervals for masses for the best-fit template ($\log(g) = 3.5$ and $T = 22000$ K). The panel on the left shows the PDF for M_X marginalized over M_{opt}. Similarly, the lower panel shows the PDF for M_{opt}, marginalized over M_X. In these panels, the the solid, dashed and dotted lines show the peak and 68.3%, 95.4% confidence intervals respectively. We obtain $M_X = 2.7^{+0.7}_{-0.6}$ $M_\odot$, and $M_{opt} = 18.1^{+2.0}_{-1.9}$ $M_\odot$. For completeness, fits for the less likely scenario with $\log(g) = 4.0$ and $T = 23000$ K are shown in blue. In this case, $M_X = 2.5 \pm 0.6$ $M_\odot$, $M_{opt} = 16.1^{+1.8}_{-1.7}$ $M_\odot$.

To test the validity of this technique, we apply it to two well-studied targets: LMC X-4 and SMC X-1. We find that the basic application of our method is overestimating the mass (Table 6.3). Part of this discrepancy is likely related to equating the Roche volume radius to the effective photometric radius (Equation (6.7)). The Roche volume radius (Equation (6.5)) is the radius of a sphere with the same volume as the Roche lobe of the star. The effective photometric radius (Equation (6.1)) is the radius of a sphere with the same surface area as the star. Since an ellipsoid has a larger surface area than a sphere of the same volume, the actual Roche volume radius will be smaller than the effective photometric radius. Using a larger radius for the Roche lobe increases the masses of both the components in the binary.

Putting it another way, if we use the Roche volume radius to calculate the brightness of the star, we will get a number lower than the observed brightness. A similar discrepancy is observed by Massey et al. (2012) in massive binaries in the LMC. They find that the absolute magnitude of LMC 172231 calculated using

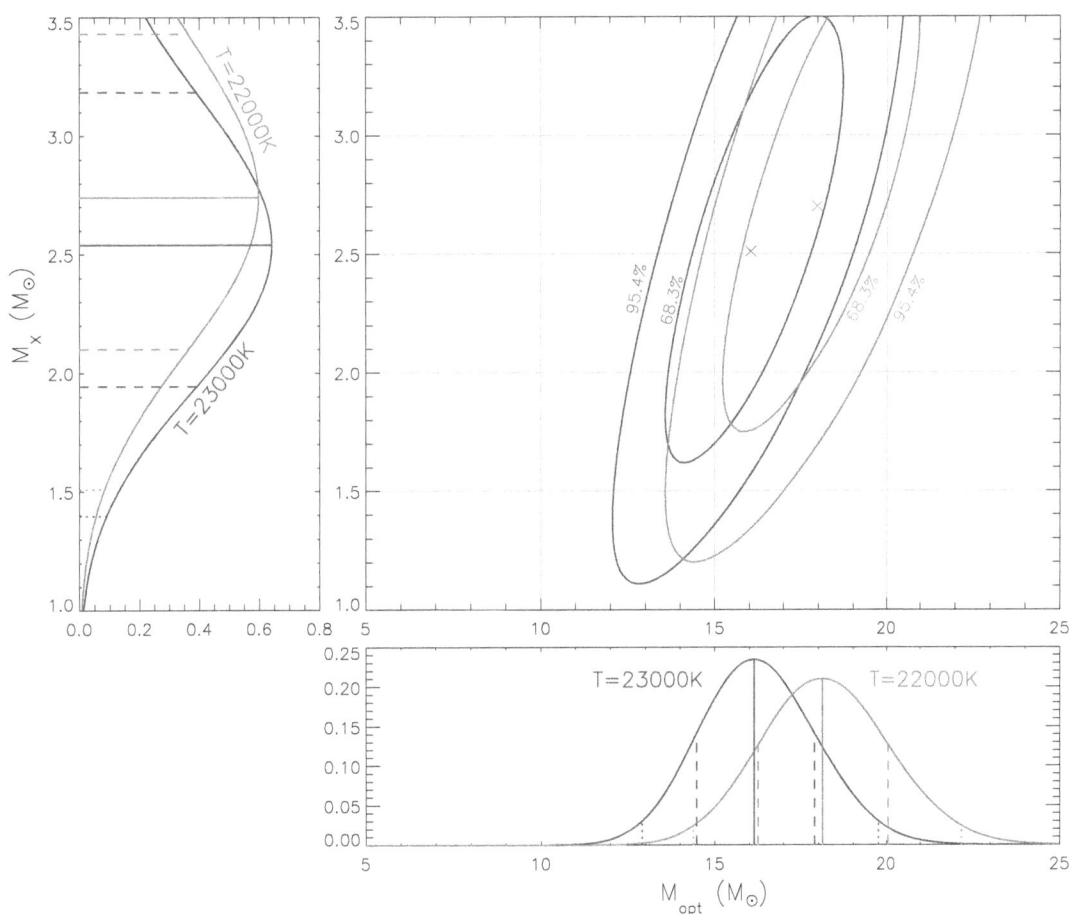

Figure 6.9. Probability density plot for neutron star mass (M_X) as a function of OB star mass (M_{opt}). Red contours show the 68.3%, 95.4% and 99.7% confidence intervals for masses using the best-fit template ($\log(g) = 3.5$ and $T = 22000$ K). The panel on the left shows the PDF for M_X marginalized over M_{opt}. Similarly, the lower panel shows the PDF for M_{opt}, marginalize over M_X. In these panels, the the solid, dashed and dotted lines show the peak and 68.3%, 95.4% confidence intervals respectively. We obtain $M_X = 2.61^{+0.32}_{-0.29}$ M$_\odot$, $M_{opt} = 18.3^{+1.7}_{-1.6}$ M$_\odot$, and $M_X > 2.0$ M$_\odot$ with 98.6% probability. Contours for the the less likely scenario with $\log(g) = 4.0$ and $T = 23000$ K are shown in blue. In this case, $M_X = 2.43^{+0.30}_{-0.28}$ M$_\odot$, $M_{opt} = 16.4^{+1.5}_{-1.4}$ M$_\odot$, and $M_X > 2.0$ M$_\odot$ with 95.1% probability.

Table 6.3. System parameters for SMC X-1 and LMC X-4

Property	SMC X-1	LMC X-4
Period (P)	3.89 d	1.41 d
P_{spin}	0.708 s	13.5 s
$a_X \sin i$ (lt-s)	53.4876 $\pm$0.0004	26.343 $\pm$0.016
Eclipse half-angle (θ_e)	26° − 30.5°	27° $\pm$2°
Mean systemic velocity (γ_{opt})	−191 $\pm$6 km s^{-1}	306 $\pm$10 km s^{-1}
Velocity semi-amplitude (k_{opt})	20.2 $\pm$1.1 km s^{-1}	35.1 $\pm$1.5 km s^{-1}
Companion spectral type	B0I	O8III
Companion T_{eff}	29000 K	35000 K
Distance (d)	60.6 $\pm$1 kpc[a]	49.4 $\pm$1 kpc[b]
Visual magnitude (m_V)[c]	13.3 $\pm$0.1	14.0 $\pm$0.1
Extinction (A_V)[d]	0.12 $\pm$0.01	0.25 $\pm$0.04
NS mass (M_X)		
van der Meer et al. (2007)	1.06 $^{+0.11}_{-0.10}$ M$_\odot$	1.25 $^{+0.11}_{-0.10}$ M$_\odot$
Our calculation	1.32 $^{+0.16}_{-0.14}$ M$_\odot$	2.05 $^{+0.24}_{-0.23}$ M$_\odot$
OB star mass (M_{opt})		
van der Meer et al. (2007)	$\approx$ 15.7 M$_\odot$	$\approx$ 14.5 M$_\odot$
Our calculation	21.1 $^{+3.2}_{-2.9}$ M$_\odot$	29.1 $^{+4.8}_{-4.4}$ M$_\odot$

Note. — Data from van der Meer et al. (2007).

[a]Hilditch et al. (2005).

[b]Freedman et al. (2001).

[c]Conservative 0.1 mag errors assumed.

[d]Foreground extinction to galaxy only. Error bars are approximate, the dominant uncertainty has been assigned to m_V.

a spherical approximation is 0.45 mag fainter than observed, while for the triple system [ST92]2-28, the numbers are consistent within errors. Using system parameters derived by van der Meer et al. (2007), we find a similar offset of 0.45 mag for LMC X-4 and 0.2 mag for SMC X-1. If we incorporate this uncertainty by allowing offsets of 0–0.4 mag, we get $M_X = 2.2^{+0.7}_{-0.6}$ $M_\odot$ and $M_{opt} = 13 \pm 4$ $M_\odot$.

Thus while this method has potential, more detailed modeling of the primary is clearly required to accurately infer component masses.

6.5 Conclusion

From our spectroscopic measurements we find that the donor star in XMMU J013236.7+303228 is a B1.5IV sub-giant with effective temperature $T = 22000 - 23000$ K. The higher temperature, $T = 33000$ K reported by Pietsch et al. (2009) is inconsistent with the absence of He II lines at, e.g., 4541 and 4686 Å in our spectra. Assuming a circular orbit, we measure a mean systemic velocity $\gamma_{opt} = -80 \pm 5$ km s^{-1} and velocity semi-amplitude $K_{opt} = 63 \pm 12$ km s^{-1} for the B star. M33 is nearly face-on, with recession velocity of -179 km s^{-1} (de Vaucouleurs et al., 1991) - so this binary seems to be moving away from the disc at 100 km s^{-1}.

Using the physical properties of the B star determined from our optical spectroscopy we find a mass for the donor of $M_{opt} = 11$ $M_\odot$. This mass is based on stellar evolution models, and will be reasonably accurate so long as binary evolution has not significantly altered the mass-luminosity relation. However, it is difficult to test this assumption based on any available observations. Using this spectroscopic mass, we calculate the mass of the compact object, $M_X = 2.0 \pm 0.4$ $M_\odot$. This is higher than the canonical 1.4 $M_\odot$ for neutron stars, but comparable to masses of other neutron stars in X-ray binaries such as the HMXB Vela X−1 (1.88 ± 0.13 $M_\odot$; Barziv et al., 2001; Quaintrell et al., 2003) or the Low Mass X-ray binaries Cyg X−2 (1.71 ± 0.21 $M_\odot$; Casares et al., 2010) and 4U 1822−371 (1.96 ± 0.35 $M_\odot$; MunozDarias et al., 2005). Since no pulsations have been detected we have only indirect evidence, based on the X-ray spectrum, that the compact object is a neutron star. However, the mass we derive here is smaller than would be expected for a black hole.

Based on the stable X-ray flux, we infer that the donor is transferring mass to the neutron star by Roche lobe overflow. By equating the Roche lobe radius to physical radius of $R = 9.1(8.7) \pm 0.3$ $R_\odot$, derived from the known distance to M33, combined with the stellar luminosity and temperature, we derive an additional orbital constraint. From a first pass calculation with a spherical approximation for the shape of the primary we get $M_X = 2.7^{+0.7}_{-0.6}$ $M_\odot$ and $M_{opt} = 18.1^{+2.0}_{-1.9}$ $M_\odot$. However, applying this technique to the well-studied binaries LMX X-4 and SMC X-1, both of which have measured component masses, we find this technique consistently overestimates the compact object mass. This is likely because the Roche surface is not spherical but elongated, which is not taken into account in our calculation. Future efforts to more accurately model the system geometry will improve the accuracy of this technique, which is applicable to Roche lobe overflow systems with known distances.

Acknowledgments

We thank Brian Grefenstette for helping with timing analyses of XMM data.

Some of the data presented herein were obtained at the W.M. Keck Observatory, which is operated as a scientific partnership among the California Institute of Technology, the University of California and the National Aeronautics and Space Administration. The Observatory was made possible by the generous financial support of the W.M. Keck Foundation.

This research has made use of the NASA/IPAC Extragalactic Database (NED) which is operated by the Jet Propulsion Laboratory, California Institute of Technology, under contract with the National Aeronautics and Space Administration. This research has made use of NASA's Astrophysics Data System Bibliographic Services. This research used the facilities of the Canadian Astronomy Data Centre operated by the National Research Council of Canada with the support of the Canadian Space Agency.

Chapter 7

The White Dwarf Companion of a 2 $M_\odot$ Neutron Star

Varun B. Bhalerao and S. R. Kulkarni

Department of Astronomy, California Institute of Technology, Pasadena, CA 91125, USA

Abstract

We report the optical discovery of the companion to the 2 $M_\odot$ millisecond pulsar PSR J1614–2230. The optical colors show that the 0.5 $M_\odot$ companion is a 2.2 Gyr old He–CO white dwarf. We infer that $\dot{M}$ during the accretion phase is $< 10^{-2}\ \dot{M}_{\rm edd}$. We show that the pulsar was born with a spin close to its current value, well below the rebirth line. The spin-down parameters, the mass of the pulsar, and the age of the system challenge the simple recycling model for the formation of millisecond pulsars.

7.1 PSR J1614–2230

PSR J1614–2230, a 3.15 ms pulsar, was discovered in a radio survey of unidentified EGRET gamma ray sources using the Parkes Radio Telescope (Hessels et al., 2005). Subsequently, X-ray emission from *Newton XMM* (Roberts et al., 2007) and γ-ray emission from *Fermi Gamma Ray Large Area Space Telescope* (Abdo et al., 2010) was detected. Like most millisecond pulsars (MSPs), PSR J1614–2230 is in a binary system. The circular orbit is consistent with the pulsar having undergone mass transfer and spun up. The mass function derived from pulsar timing indicated a companion with mass $M_2 > 0.4\ M_\odot$ (Hessels et al., 2005).

The system recently came into prominence when Demorest et al. (2010) reported the mass of the pulsar to be $1.97 \pm 0.04\ M_\odot$. The detection of such a massive neutron star (NS) places very strong constraints on the equation of state of matter at extreme nuclear densities (see, for example, Lattimer & Prakash, 2004, 2005). The rather exquisite precision of this mass measurement was possible due to the orbit being almost perpendicular to the plane of the sky. As a result, the Shapiro delay caused by the companion is very large, resulting in a precise estimate of the mass of the companion, $M_2 = 0.500 \pm 0.006\ M_\odot$. The 8.7 day orbital period is significantly shorter than ~ 120 days expected for a low-mass X-ray binary with such a massive secondary (Rappaport et al., 1995)—suggesting a peculiar evolutionary history for this binary.

A version of this chapter was published in the *Astrophysical Journal* (Bhalerao & Kulkarni, 2011). It is reproduced here with permission from AAS.

Given the importance of the result of Demorest et al. (2010) additional verification or consistency checks of physical parameters of PSR J1614–2230 can be expected to be of some value. A 0.5 M$_\odot$ white dwarf (WD) at the inferred distance of PSR J1614–2230 ($d \sim 1.2$ kpc), even if a few Gyr old, is within the reach of present-day optical telescopes. It is this search for the WD that constitutes the principal focus of this Letter.

7.2   Observations at the W. M. Keck Observatory

We observed PSR J1614–2230 (Figure 7.1) in g and R bands using the imaging mode of the Low Resolution Imaging Spectrograph (LRIS) on the 10 m Keck-I telescope (Oke et al., 1995), with upgraded red[1] and blue cameras (McCarthy et al., 1998; Steidel et al., 2004). Several images were acquired at each target location, dithering the telescope by small amounts between each exposure. The observing conditions on UT 2010 May 15 were poor (seeing $1''.4$), so only data acquired on UT 2010 July 8 (R band seeing $0''.85$ FWHM) were used in this analysis. The total exposure on this night was 960 s in the R band and 1010 s in the g band. The plate scale is $0''.135$ pixel^{-1} for both cameras

The images were processed using IRAF. After bias correction and flat fielding, cosmic rays were rejected using L.A.Cosmic (van Dokkum, 2001). The images were then aligned with xregister and averaged to produce the final image for each band (see Figure 7.1). A World Coordinate System was calculated using USNO-B1.0 stars in the field, with the imcoords package. Using 33 stars in the field, for the final R-band image we obtained an RMS error of $0''.14$ in right ascension (R.A.) and $0''.20$ in declination, adding up to a radial error of $0''.24$. For the final g-band image, the residuals were $0''.13$ and $0''.18$ for R.A. and declination, respectively, giving a total error of $0''.22$.

We measured fluxes with aperture photometry using the IDL APPHOT package. For each night, we measured the seeing (FWHM) and set the aperture to one seeing radius (Mighell, 1999). We extracted the sky from an annulus 5–10 seeing radii wide. We had observed nearby a Sloan Digital Sky Survey (SDSS; York et al., 2000) field[2] with the same settings as the science field. The calibration field was observed immediately after the science exposures, and had airmass 1.4 as compared to 1.6 for the target. We used the magnitudes of six stars from that field to calculate the photometric zero point and calibrated six reference stars in the science field (Figure 7.1, Table 7.1). R band magnitudes were calculated using photometric transformations[3] prescribed by Jester et al. (2005) for stars with $R_c - I_c < 1.15$. The typical uncertainty in g-band magnitudes for stars with $m_g \sim 20$ is 0.03 mag. In R band we have $\Delta m_R = 0.07$ for $m_R \sim 20$, including the uncertainty in the transformations.

7.3   Detection of an Optical Counterpart

In the vicinity of the nominal pulsar position, we find a faint source (labeled "P") in the R band (Figure 7.1). We do not detect anything within $1''$ of the target in the g band. The optical coordinates of this source, the timing position of the pulsar and a proposed X-ray counterpart are summarized in Table 7.2. To compare this with the source location, we first have to correct for the 33 mas yr^{-1} proper motion of the source. The LRIS source P is about $0''.50$ South of the pulsar position (extrapolated for the epoch of

---

[1]http://www2.keck.hawaii.edu/inst/lris/lris-red-upgrade-notes.html
[2]Calibration SDSS field: $\alpha = 17^{\rm h}19^{\rm m}10^{\rm s}.10$, $\delta = -14°38'46''.0$.
[3]Photometric transforms: $V = g - 0.59(g - r) - 0.01 \pm 0.01$ and $V - R = 1.09(r - i) + 0.22 \pm 0.03$.

Table 7.1. Positions and magnitudes of reference stars

ID	RA $\alpha - 16^{\mathrm{h}}14^{\mathrm{m}}$	Declination $\delta - (-22°)$	m_R[a]	m_g
A	$36^{\mathrm{s}}.98(8)$	$29'18''.78(11)$	$17.94(8)$	$18.73(3)$
B	$36^{\mathrm{s}}.29(7)$	$29'31''.97(7)$	$17.71(8)$	$18.38(3)$
C	$35^{\mathrm{s}}.88(9)$	$30'18''.70(10)$	$19.76(10)$	$20.41(4)$
D	$34^{\mathrm{s}}.39(5)$	$30'12''.11(12)$	$19.15(8)$	$21.24(4)$
E[b]	$35^{\mathrm{s}}.66$	$31'04''.17$	$21.38(9)$	$22.04(5)$
F	$34^{\mathrm{s}}.92(16)$	$30'59''.46(2)$	$20.18(9)$	$21.36(5)$
G[b]	$37^{\mathrm{s}}.50$	$30'43''.69$	$20.71(8)$	$21.87(4)$
Q	$36^{\mathrm{s}}.47(7)$	$30'35''.90(4)$	(Saturated)	$17.20(3)$
R	$35^{\mathrm{s}}.92(2)$	$30'30''.45(1)$	$20.10(8)$	$20.94(4)$
S	$36^{\mathrm{s}}.50(10)$	$30'13''.93(13)$	(Saturated)	$17.75(3)$

Note. — Stars A–G were used as reference stars for photometry. Right ascension and declination were obtained from USNO-B1.0 unless otherwise specified.

[a] R band magnitudes calculated using SDSS magnitudes and Jester et al. (2005) transformation equations (Section 7.2). The numbers in parenthesis do not include a 0.03 mag uncertainty in absolute calibration due to the transformations.

[b] Coordinates obtained from our final R-band science image. The World Coordinate System for this image was calculated using a total of 33 USNO-B1.0 stars (Section 7.2).

Table 7.2.   Coordinates of PSR J1614–2230 at different epochs

Method	Epoch	Ecliptic Longitude (λ)	Ecliptic Latitude (β)	RA $\alpha - 16^\mathrm{h}14^\mathrm{m}$	Declination $\delta - (-22°30')$
Timing	J2005.63	245.78827556(5)	$-1.256744(2)$	$36^\mathrm{s}.5051(1)$	$31''.080(7)$
Timing	J2000.00	245.78826025(12)	$-1.256697(5)$	$36^\mathrm{s}.5034(2)$	$30''.904(19)$
Timing	J2007.32	245.78828015(6)	$-1.256758(2)$	$36^\mathrm{s}.5056(1)$	$31''.132(9)$
Chandra[a]	J2007.32	$\cdots$	$\cdots$	$36^\mathrm{s}.50(15)$	$31''.33(20)$
Timing	J2010.51	245.78828886(11)	$-1.256785(5)$	$36^\mathrm{s}.5067(2)$	$31''.23(2)$
LRIS	J2010.51	$\cdots$	$\cdots$	$36^\mathrm{s}.50(16)$	$31''.72(20)$

Note. — The proper motion of the pulsar as obtained from timing observations is $\mu_\lambda = 9.79(7)$ mas yr^{-1}, $\mu_\beta = -30(3)$ mas yr^{-1}. This proper motion is used to estimate the ecliptic coordinates of the pulsar at the epoch of the *Chandra* and LRIS observations. The right ascension and declination are calculated from ecliptic coordinates using the `Euler` program in `IDL`. The equinox in all cases is J2000.

[a]Bore-sight corrected coordinates. The target and the reference source "E" are detected in the *Chandra* image. We extract source coordinates using `celldetect`. We assume that the *Chandra* coordinate system and our R-band coordinate solution are related by a simple offset with no rotation. Using the X-ray and R-band coordinates of source E, we calculate that the offset is $\alpha_X - \alpha_R = 0''.21$, $\delta_X - \delta_R = -0''.20$ and correct the target position using this offset.

LRIS observations). Given the $0''.24$ (1-σ) astrometric uncertainty of the optical images, this position is consistent with the location of the pulsar.

The density of objects brighter than star P in this field is 0.02 arcsecond^{-2}. Using the $0''.85$ seeing FWHM as the mean diameter of stars, we calculate a false identification probability of 1%. The excellent astrometric coincidence and the low probability of chance coincidence embolden us to suggest that star P is the optical counterpart of PSR J1614–2230.

Counterpart P is located only $4''.2$ from the 16.3 mag star USNO-B1.0 0674-0429635, and is contaminated by the flux in the wings of its point-spread function. The proximity to this bright source will bias both the photometry and the astrometry of the counterpart. To calculate the bias, we injected fake Gaussian sources with FWHM matched to the seeing and brightness comparable to the faint object. We then measured the coordinates and magnitude of the injected source using the same procedure as for the faint object. We find that for separations $\approx 4''.2$, the recovered position is systematically pulled towards the bright star by $0''.1$–$0''.2$. This is small enough that we do not apply this correction. The same exercise led us to derive the photometric bias. The de-biased R-band magnitude of P is $m_R = 24.3 \pm 0.1$.

To infer properties of the WD, we need to calculate its absolute magnitude. To calculate the optical extinction, we assume that the ratio N_H/N_e is constant along the line of sight in the direction of PSR J1614–2230. The dispersion measure (DM) for PSR J1614–2230 is 34.5 pc cm^{-3} (Demorest et al., 2010), and the total DM in this direction is 104 pc cm^{-3} (Cordes & Lazio, 2002)—about a factor of three higher than the pulsar DM. This implies that the pulsar is behind approximately one third of the galactic absorbing column. We can then scale the total R band extinction in this direction ($A_{R,\mathrm{total}} = 0.65$; Schlegel et al., 1998) to get $A_R = 0.22$. We assume $\lambda_{\mathrm{eff}} = 0.47$ μm for the g band[4], and use the standard reddening law with $R_V = 3.1$ to get $A_g = 0.34$ (Cox, 2000).

Demorest et al. (2010) use the DM to estimate that PSR J1614–2230 is at a distance $d = 1.2$ kpc. For this distance, the extinction-corrected R-band absolute magnitude is $M_R = 13.7 \pm 0.1$. Furthermore, they place a lower limit of 900 pc on the distance, using pulsar parallax from timing measurements. The corresponding absolute magnitude is $M_R = 14.3 \pm 0.1$.

7.4 Pulsar age and birth spin period

The measured mass (0.5 M$_\odot$) and the inferred absolute magnitude ($M_R \approx 13.7$) when applied to WD cooling models (Chabrier et al., 2000) lead to an age of $\tau_{\mathrm{WD}} \sim 2.2$ Gyr. For such a WD, the expected absolute magnitude in other bands is $M_B = 14.6$ and $M_V = 14.1$. This gives $m_g \approx 25.0$, consistent with our non detection.

As per the standard evolutionary model for MSPs, the NS in PSR J1614–2230 was spun up by accretion from a low- or intermediate-mass companion star. The accretion stopped when the companion decoupled from its Roche lobe and became a WD. At this point, the MSP started spinning down by radiating energy. The spin-down age of the MSP is thus equal to the cooling down age of the WD.

The period of the pulsar at birth (P_{init}), its spin-down age (τ), present-day period (P) and period derivative ($\dot{P}$) are related to each other as follows:

$$\tau = \frac{P}{(n-1)\dot{P}} \left[1 - \left(\frac{P_{\mathrm{init}}}{P}\right)^{n-1}\right] \tag{7.1}$$

---

[4]http://www.sdss.org/dr6/instruments/imager/index.html#filters

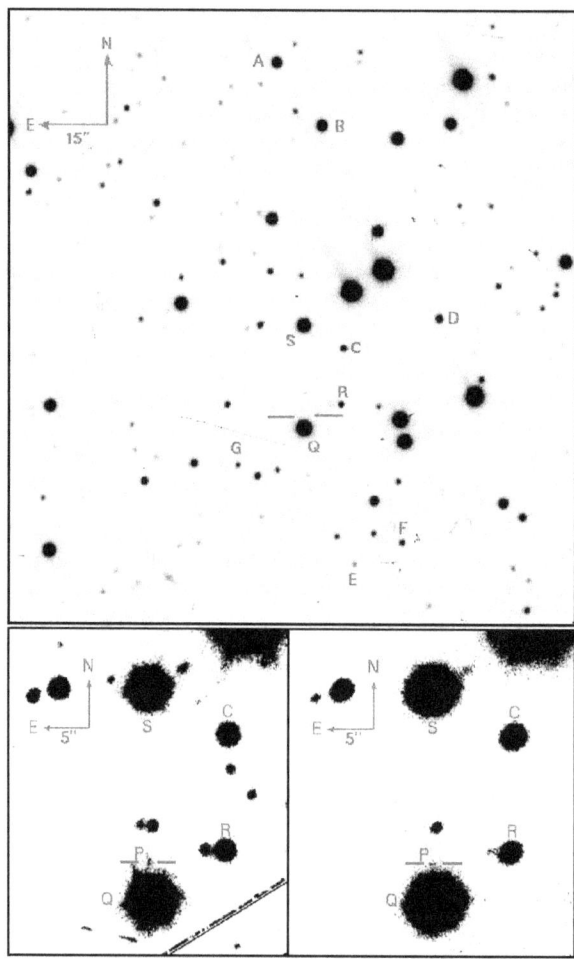

Figure 7.1. Top: R band LRIS image of the PSR J1614–2230 field. The expected location of the target for epoch 2010.52 is shown by two horizontal lines just below the center of the image. The CCD shows considerable blooming in rows and columns around bright stars. The hexagonal mirror shape gives six diffraction spikes separated by 60°. To avoid contamination of the target by these artifacts from the bright star Q, we set the position angle to 300°. Bottom: R-band (left) and g-band (right) images of PSR J1614–2230. The target (P) is detected in the R band at $\alpha = 16^h14^m36^s.50$, $\delta = -22°30'31''.72$ and is marked with thick horizontal lines. There is a $0''.33$ offset between the expected and observed positions of the target (Table 7.2). The target is not detected in the g-band, but the R band location is marked for reference. The diagonal streak in the R image is a bad CCD column.

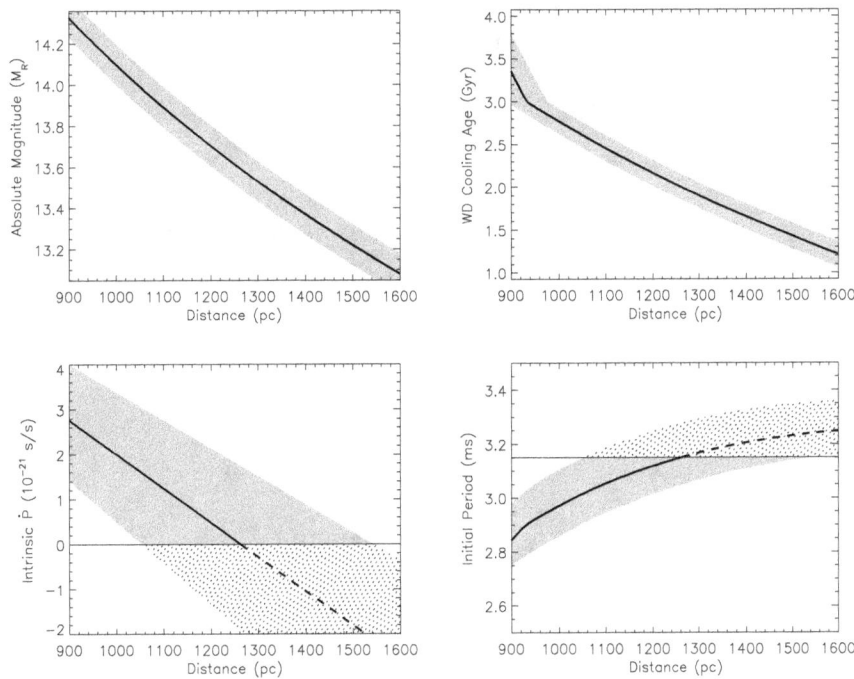

Figure 7.2. Inferred parameters for PSR J1614–2230 as a function of distance. The gray or dotted regions show the error bars on each parameter. Top left: The absolute R-band magnitude of the WD. Top right: WD cooling age inferred from Chabrier et al. (2000). The kink at 3 Gyr may be a result of granularity of the tables. Bottom left: the intrinsic period derivative ($\dot{P}$) in the pulsar frame, corrected for the Shklovskii effect. Since the pulsar cannot be spinning up, the values of $\dot{P} < 0$ are unphysical and are shown as the dotted area. The maximum distance to the pulsar is inferred to be 1540 pc. Lower right: the initial spin period of the pulsar. Birth periods slower than the current period are excluded.

where n is the "braking index" for the pulsar, with $n = 3$ appropriate for a dipole radiating into vacuum. Thus, the period at birth is given by

$$P_{\text{init}} = P \left[1 - \frac{\tau \dot{P}(n-1)}{P} \right]^{1/(n-1)} \tag{7.2}$$

The measured period derivative of a pulsar ($\dot{P}_{\text{obs}}$) is always higher than its true period derivative ($\dot{P}$) owing to transverse motion (Shklovskii, 1970). The corrected period derivative is $\dot{P} = \dot{P}_{\text{obs}} - P\mu^2 d/c$, where d is the distance and μ is the proper motion. Demorest et al. (2010) measure $\mu = 32(3)$ mas yr^{-1} for PSR J1614–2230. Using the nominal distance $d = 1.2$ kpc, $\dot{P} = 4.8 \times 10^{-22}$ s s^{-1}.

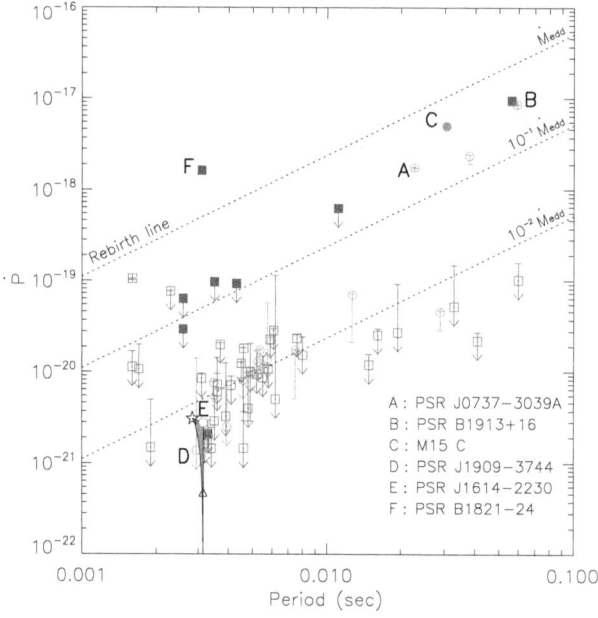

Figure 7.3. Pulsar P–$\dot{P}$ diagram. Period derivatives are corrected for the Shklovskii effect. Some interesting systems are labeled. Filled symbols are pulsars in globular clusters, whose $\dot{P}$ may be affected by the dense environment. Circles denote NS binaries with mass measurements. The dashed lines are the pulsar rebirth lines $\dot{P} = (\dot{M}/\dot{M}_{\text{edd}})1.1 \times 10^{-15}$ s$^{-4/3}P^{4/3}$ (Arzoumanian et al., 1999) for $\dot{M}/\dot{M}_{\text{edd}} = 1$, 10^{-1} and 10^{-2}. The location of PSR J1614–2230 on this plot depends on the distance (see Section 7.4). Were the pulsar to be at 900 pc, it would be born at the star symbol and evolve toward its present-day location, the hollow circle marked "E". If instead it is at 1200 pc, the birth location is shown by the upward triangle, which also coincides with its present-day location. The solid curve passing through the star and the upward triangle denotes all possible birth locations for PSR J1614–2230. The system evolves through the shaded gray area to its present-day location on the vertical line through E and the upward triangle.

The DM inferred distance is quite uncertain, so it is useful to consider the dependence of all parameters on distance. Figure 7.2 shows the range of values for the R-band absolute magnitude (M_R), the WD cooling age (τ_{WD}), the intrinsic period derivative in the pulsar's frame ($\dot{P}$) and the initial spin period (P_{init}). Since there is no energy injection to the pulsar now, it must be currently spinning down. This implies $\dot{P} \geq 0$ and allows us to calculate an upper limit on the distance to the pulsar: $d_{max} = 1540$ pc (1-σ). The last panel in Figure 7.2 shows $P_{init} \geq 2.75$ ms. We conclude that the pulsar in PSR J1614–2230 must have been born with a period close to the current observed value.

The birth spin period of a NS is governed by an equilibrium between the ram pressure of the accreting material and the magnetic field. NSs spun up by accretion at the Eddington rate ($\dot{M}_{edd}$) are reborn as MSPs on the "rebirth line" (Arzoumanian et al., 1999). Figure 7.3 shows this rebirth line on a P–$\dot{P}$ diagram, along with current positions of various pulsars from the ATNF database (Manchester et al., 2005). Also shown are rebirth lines for pulsars accreting at $10^{-1}\dot{M}_{edd}$ and $10^{-2}\dot{M}_{edd}$. For a pulsar radiating as a dipole, the $P\dot{P}$ product remains constant through its lifetime. Thus, we can calculate the initial spin-down rate of the pulsar: $\dot{P}_{init} = P\dot{P}/P_{init}$, where P_{init} comes from Equation 7.2. If PSR J1614–2230 is at 900 pc, then it would have been born with $\dot{P}_{init} = 3.1 \times 10^{-21}$ s s^{-1} and $P_{init} = 2.84$ ms. This value is indicated with a star. The pulsar then evolves toward the lower right, to the circle marked "E". Similarly, the birth parameters for the pulsar assuming $d = 1.2$ kpc are shown by the upward triangle—it is nearly coincident with the current parameters of the pulsar for this distance. Other possible birth locations for the pulsar lie along the solid line passing through the star and triangle. The gray region shows all possible positions that PSR J1614–2230 can have occupied in its lifetime. It is clear that the pulsar was born well below the rebirth line, and the mean accretion rate during the spin-up phase (the final major accretion phase) was lower than $10^{-2}\dot{M}_{edd}$.

Two groups have run detailed simulations of the evolution of PSR J1614–2230. The Lin et al. (2011) model and the preferred model of Tauris et al. (2011) are qualitatively similar: the system begins as an intermediate mass X-ray binary consisting of a NS and a ~ 4 M$_\odot$ main-sequence secondary, which evolves to form a CO WD with an He envelope. The secondary accretes mass onto the NS in three phases. The first phase (A1) is a thermal timescale mass transfer at super-Eddington accretion rates, where the NS gains little mass. The next phase (A2) is on a nuclear timescale (~ 35 Myr), when the secondary is burning H in the core and envelope. During this phase, the accretion rates are upto about a tenth of the Eddington limit. During the final accretion phase (phase AB), the secondary is burning He in its core and H in an envelope. This causes the radius of the donor star to expand, triggering accretion at near-Eddington rates for 5–10 Myr. The NS gains the most mass during this phase. For typical NS parameters, accreting 0.1–0.2 M$_\odot$ is enough to spin them up millisecond periods (Kiziltan & Thorsett, 2010). Hence, the near-Eddington accretion in phase AB should spin the pulsar up all the way to the rebirth line. This is inconsistent with our inferred birth position of PSR J1614–2230.

If the stellar evolution models are correct, then there is a problem with the standard formation scenario of MSPs (Radhakrishnan & Srinivasan, 1982; Alpar et al., 1982). In essence, the birth period depends on factors other than the magnetic field strength and accretion rates. For instance, Bildsten (1998, 2003) proposes that accretion induces a quadrupole moment Q in the NS. The NS then loses angular momentum by gravitational wave radiation. Since the magnetic fields do not play any significant role in this model, the rebirth line becomes irrelevant, and PSR J1614–2230 may start its life anywhere on the P–$\dot{P}$ diagram. In summary, the birth of PSR J1614–2230 away from the rebirth line requires reconsideration of angular momentum loss mechanisms.

Going forward, on-going radio observations (pulsar timing and VLBI) should improve the parallax and thereby decrease the uncertainty in the inferred $\dot{P}$ and thus better locate the pulsar in the P-$\dot{P}$ diagram. The scattered light from light Q, while bothersome to the present observations, provide an opportunity to use adaptive optics to measure the near-IR fluxes of the WD. Grism spectroscopy with

Hubble Space Telescope can potentially reveal the presumed H+He layer posited by stellar evolutionary models of Tauris et al. (2011).

Acknowledgments. We are grateful to Scott Ransom for providing the source position in advance of the publication. We thank M. Kasliwal and A. Gal-Yam for undertaking the observations.

Some of the data presented herein were obtained at the W.M. Keck Observatory, which is operated as a scientific partnership among the California Institute of Technology, the University of California and the National Aeronautics and Space Administration. The Observatory was made possible by the generous financial support of the W.M. Keck Foundation.

Chapter 8

Conclusions and Future Work

When we started the X-Mas survey, masses had been measured for only six neutron stars in high-mass X-ray binaries. To date, we have contributed three of the dozen known masses in this list (Figure 8.1). Our measurements fully support the initial hypothesis that NSs in HMXBs have a wide spread in masses. X-Mas is an ongoing program, and we will add a few more measurements to this list.

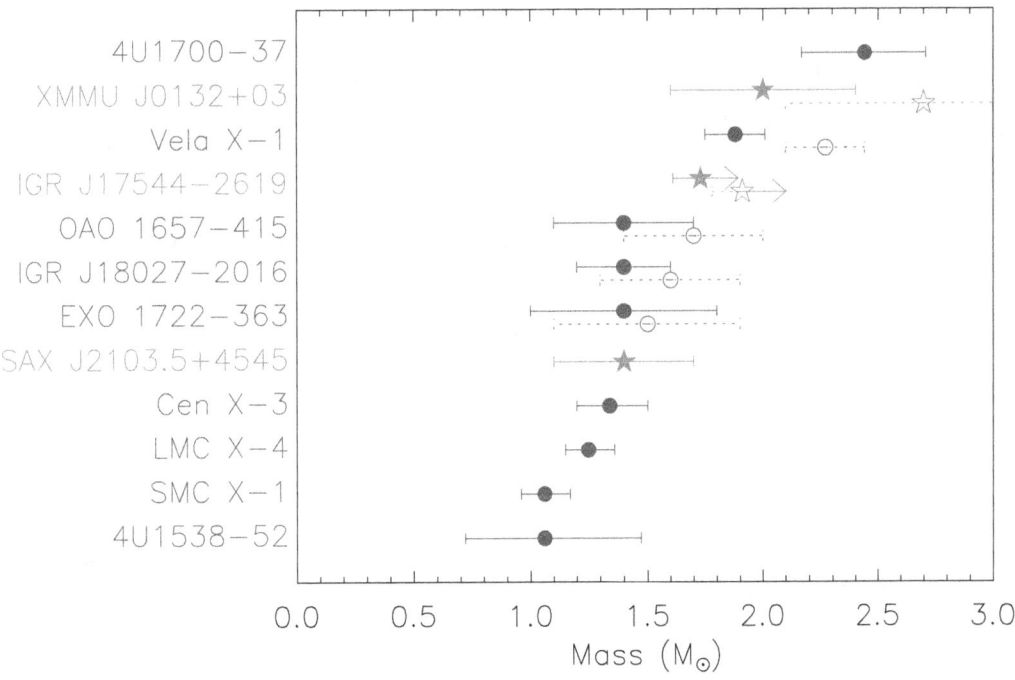

Figure 8.1. All known masses of neutron stars in HMXBs. Red ★ indicate mass measurements discussed in this thesis, blue ● are measurements in literature. For Vela X-1, OAO 1654–415, IGR J18027–2016 and EXO1722–363, the solid and hollow symbols correspond edge-on and Roche-filling systems respectively. For references see Table 4.1.

Neutron star mass measurements have come to a stage where they provide critical feedback to fundamental physics. The recent discovery of a two solar mass neutron star (Demorest et al., 2010) has ruled out some equations of state which preclude the existence of such heavy NSs (Figure 4.1). We can push these limits futher by finding heavier NSs or improving constraints on the existing heavy candidates like 4U 1700−37 (Clark et al., 2002) and XMMU J013236.7+303228 (Chapter 6). For instance, the discovery of a 2.5 $M_{\odot}$ object can rule out most proposed EOSs, leaving only a few candidates like MS0 and MS2 (Lattimer & Prakash, 2007).

Beyond constraints on the EOS, an intriguing open question is the paucity of compact objects in the 2.5 − 6 solar mass regime (e.g., Remillard & McClintock, 2006; Özel et al., 2010). While all neutron stars have masses $\lesssim 2.5\ M_{\odot}$, the six stellar-mass black holes with dynamical mass measurements are all heavier than 6 $M_{\odot}$ (Figure 8.2). This gap in masses of compact objects persists even after accounting for lower limits on masses of other black holes (Özel et al., 2012). While fundamental physics sets the maximum mass for a NS, it does not rule out the formation of, say, a four solar mass black hole. Instead, this gap must arise from the astrophysical scenarios in which black holes form. This gap also suggests the possibility that the distribution of NS masses will probably not have a sharp cutoff governed by fundamental physics. Instead, the formation mechanisms pertaining to core collapse of massive stars do not form compact objects in this regime. More and more robust neutron star mass measurements will enable us to confirm or refute the existence of this gap.

Over the course of this project, we have developed and refined methods for obtaining accurate radial velocity measurements of OB stars. We learned some important lessons along the way, which will help us in planning any future observations. For planning the observing runs, wherever possible, we will choose the spectrograph position angle to place a reference star on the slit along with the target. We can measure the velocity of the target with respect to this star, to overcome systematic errors in data analysis. We will also consider using multi-slit spectrographs for this purpose. In previous observing runs, we obtained calibration arc data at the position of each target. In data analysis we discovered that DBSP and LRIS have very stable wavelength solutions, so we can save time by skipping this step, and obtaining more arc lamp exposures and flat fields at the start of the night. It was useful to split the total observing time for a target into multiple exposures. In case there were problems like tracking issues or bad readout during any of the exposures, it does not ruin the complete data for that target for that night. We also learned several lessons in data reduction and analysis, which often involved reprocessing data to get better results. I have compiled these in a separate document, available online at http://authors.library.caltech.edu/31417.

This field is in an exciting phase now, with several developments favoring the exploration of NS masses. *Integral* is continuing to finding new X-ray binaries. The upcoming *AstroSat* and NuSTAR missions will enable X-ray timing measurements for many such systems. Beyond X-ray binaries, extensive radio surveys are studying neutron stars for projects like Nanograv. Serendipitously discovered NS binaries in such systems can also be studied to measure NS masses. This ever increasing sample of neutron star masses will help us better understand neutron stars from the context of both: fundamental physics and astrophysics.

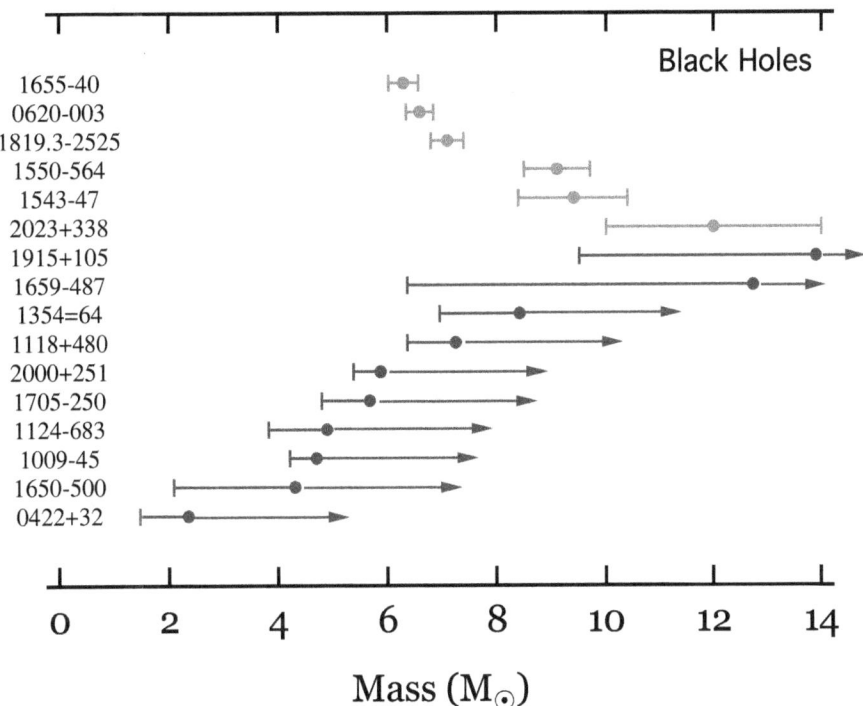

Figure 8.2. Masses of Black Holes: Dynamical measurements have yielded masses for only six stellar mass BHs (magenta data points, top). Lower limits on mass are known for several other systems (blue data points). There are no known compact objects with masses in the range of three to six solar masses. Image reproduced with permission from Özel et al. (2012).

Appendix A

The Polar Catalysmic Variable 1RXS J173006.4+033813

VARUN B. BHALERAO[a], MARTEN H. VAN KERKWIJK[a,b], FIONA A. HARRISON[c], MANSI M. KASLIWAL[a], S. R. KULKARNI[a],VIKRAM R. RANA[c]

[a]Department of Astronomy, California Institute of Technology, Pasadena, CA 91125, USA

[b]On sabbatical leave from Department of Astronomy and Astrophysics, University of Toronto, 50St. George Street, Toronto, ON M5S 3H4, Canada.

[c]Space Radiation Laboratory, California Institute of Technology, Pasadena, CA 91125, USA

Abstract

We report the discovery of 1RXS J173006.4+033813, a polar cataclysmic variable with a period of 120.21 min. The white dwarf primary has a magnetic field of $B = 42^{+6}_{-5}$ MG, and the secondary is a M3 dwarf. The system shows highly symmetric double peaked photometric modulation in the active state as well as in quiescence. These arise from a combination of cyclotron beaming and ellipsoidal modulation. The projected orbital velocity of the secondary is $K_2 = 390 \pm 4$ km s^{-1}. We place an upper limit of 830 ± 65 pc on the distance.

Keywords: binaries: close—binaries: spectroscopic—novae, cataclysmic variables—stars: individiual (1RXS J173006.4+033813)—stars: variables: other

A.1 Introduction

Cataclysmic Variables (CVs) are close interacting binary systems in which a white dwarf (WD) accretes material from a Roche lobe filling late-type secondary star (Warner, 1995; Hellier, 2001). In most non-magnetic CVs ($B < 10^4$ G), the material lost from the secondary does not directly fall onto the WD because of its large specific orbital momentum: instead, it settles down in an accretion disc around the WD.

The accretion disc is the brightest component of the CV due to the large gravitational energy release in viscous accretion. The disc dominates the emission from the WD and donor over a wide wavelength range.

A version of this chapter was published in the *Astrophysical Journal* (Bhalerao et al., 2010). It is reproduced here with permission from AAS.

113

On the other hand, the accretion geometry in magnetic CVs is strongly influenced by the WD magnetic field. Magnetic CVs are broadly divided into two subclasses: Polars and Intermediate Polars (IPs). Polars usually show a synchronous or near synchronous rotation of WD with the orbital motion of the binary system and have high magnetic fields ($B > 10$ MG) (for a review, see Cropper, 1990). In IPs the WD rotation is far from synchronous and typically have magnetic field, $B < 10$ MG (for a review, see Hellier, 2002). The strong magnetic field in polars deflects the accretion material from a ballistic trajectory before an accretion disc can form, channeling it to the WD magnetic pole(s). The infalling material forms a shock near the WD surface, which produces radiation from X-rays to infrared wavelengths. Electrons in the ionized shocked region spiral around the magnetic field lines and emit strongly polarized cyclotron radiation at optical and infrared wavelengths. Polars exhibit X-ray on (high) and off (low) states more frequently than the other variety of CVs (Ramsay et al., 2004).

1RXS J173006.4+033813 (hereafter 1RXS J1730+03) is a Galactic source that is highly variable in the optical and X-ray, exhibiting dramatic outbursts of more than 3 magnitudes in optical. It was discovered by the ROSAT satellite during its all-sky survey (Voges et al., 1999). Denisenko et al. (2009), in the course of their investigation of poorly studied ROSAT sources, reported that USNO-B1.0 object 0936-00303814 which is within the $10''$ (radius) localization of the X-ray source showed great variability (ΔR of up to 3 mag) in archival data (Palomar Sky Survey; SkyMorph/NEAT). During certain epochs the source appears to have been undetectable ($m_R > 20$ mag). Denisenko et al. (2009) undertook observations with Kazan State University's 30-cm robotic telescope and found variability on rapid timescales of 10 minutes.

In this paper, we report the results of our photometric, spectroscopic and X-ray follow-up of 1RXS J1730+03.

A.2   Observations

A.2.1   Optical Photometry

We observed 1RXS J1730+03 with the Palomar Robotic 60-inch telescope (P60; Cenko et al., 2006) from UT 2009 April 17 to UT 2009 June 5, and with the Large Format Camera (LFC; Simcoe et al., 2000) at the 5 m Hale telescope at Palomar on UT 2009 August 26. Here we give details of the photometry.

We define a photometric epoch as observations from a single night when the source could be observed. We obtained 28 epochs with the P60, subject to scheduling and weather constraints. A typical epoch consists of consecutive 90–120 s exposures spanning between 30–300 minutes (Table A.1). We obtained g', r', i' photometry on the first and third epochs. After the third epoch, we continued monitoring the source only in i' band.

We reduced the raw images using the default P60 image analysis pipeline. LFC images were reduced in IRAF[1]. We performed photometry using the IDL[2] DAOPHOT package (Landsman, 1993). Fluxes of the target and reference stars (Figure A.1) were extracted using the APER routine. For aperture photometry, the extraction region was set to one seeing radius, as recommended by Mighell (1999). The sky background was extracted from an annular region 5–15 seeing radii wide. We used flux zero points and seeing values output by the P60 analysis pipeline. Magnitudes for the reference stars were calculated from a few images. The magnitude of 1RXS J1730+03 was calculated relative to the mean magnitude of a 9 reference stars for LFC images, and 15 reference stars for P60 images (Table A.2). The LFC images (Figure A.1) resolve out a faint nearby star ($m_{i'} = 20.8$), $3''.4$ from the target. The median seeing in P60 data is $2''.1$ (Gaussian FWHM): so there is a slight contribution from the flux of this star to photometry of 1RXS J1730+03. We

---

[1]http://iraf.noao.edu/
[2]http://www.ittvis.com/ProductServices/IDL.aspx

Table A.1.   Photometry of 1RXS J1730+03

Date (UT)	HJD	Filter name[a]	Exposure time (sec)	Magnitude	Error[b]
20090417	54938.832278	g'	120	20.8	0.13
20090417	54938.835488	i'	90	19.39	0.05
20090417	54938.838696	g'	90	20.75	0.13
20090417	54938.839953	i'	90	19.17	0.05
20090417	54938.841212	g'	90	20.67	0.13
20090417	54938.842469	i'	90	19.1	0.05
20090417	54938.843727	g'	90	20.63	0.12
$\cdots$	$\cdots$	$\cdots$	$\cdots$	$\cdots$	$\cdots$
20090826	55069.691401	$lfci'$	60	20.4	0.05
20090826	55069.693218	$lfci'$	60	20.04	0.06

Note. — This table is available in a machine readable form online. A part of the table is reproduced here for demonstrating the form and content of the table.

[a]Filters g', r' and i' denote data acquired at P60 in the respective filters, $lfci'$ denotes data acquired in the i' band with the Large Format Camera at the Palomar 200" Hale telescope

[b]Relative photometry error. Values do not include an absolute photometry uncertainty of 0.16 mag in the g' band, 0.14 mag in the r' band and 0.06 mag in the i' band. Absolute photometry is derived from default P60 zero point calibrations.

Table A.2. Photometry of reference stars for 1RXS J1730+03

Identifier[a]	Right Ascension	Declination	g' magnitude	r' magnitude	i' magnitude
A	262:30:21.99	03:38:37.5	15.861 ± 0.003	15.600 ± 0.003	15.341 ± 0.003
B	262:30:14.57	03:37:11.0	17.372 ± 0.009	17.153 ± 0.006	16.872 ± 0.008
C, 10	262:32:45.57	03:37:17.0	16.879 ± 0.006	16.691 ± 0.005	16.423 ± 0.006
D, 9	262:32:13.84	03:38:03.0	18.161 ± 0.016	17.929 ± 0.010	17.660 ± 0.013
E	262:31:00.31	03:38:31.3	18.518 ± 0.020	17.462 ± 0.007	16.811 ± 0.007
F	262:30:55.06	03:37:26.4	18.530 ± 0.022	18.158 ± 0.012	17.816 ± 0.016
G, 8	262:31:57.98	03:38:03.9	18.621 ± 0.022	18.259 ± 0.012	17.905 ± 0.016
H	262:30:41.22	03:38:32.8	16.770 ± 0.006	16.081 ± 0.004	15.688 ± 0.004
I	262:29:45.31	03:38:43.7	17.661 ± 0.011	17.302 ± 0.007	16.969 ± 0.008
1	262:32:01.97	03:40:28.3	$\cdots$	$\cdots$	17.059 ± 0.035
2	262:31:44.94	03:39:48.6	$\cdots$	$\cdots$	19.855 ± 0.043
3	262:32:31.60	03:39:32.4	$\cdots$	$\cdots$	19.148 ± 0.039
4	262:32:33.94	03:38:59.4	$\cdots$	$\cdots$	17.622 ± 0.033
5	262:32:11.15	03:38:39.0	$\cdots$	$\cdots$	20.777 ± 0.049
6	262:31:35.94	03:38:29.5	$\cdots$	$\cdots$	19.843 ± 0.026
7	262:31:07.21	03:37:57.9	$\cdots$	$\cdots$	19.057 ± 0.026
11	262:31:55.09	03:36:45.4	$\cdots$	$\cdots$	17.471 ± 0.030
12	262:31:40.33	03:36:34.5	$\cdots$	$\cdots$	16.339 ± 0.044
13	262:33:19.44	03:36:21.2	$\cdots$	$\cdots$	18.006 ± 0.030
14	262:31:45.44	03:36:12.4	$\cdots$	$\cdots$	18.085 ± 0.048
15	262:32:18.64	03:35:48.1	$\cdots$	$\cdots$	17.392 ± 0.055

Note. — This table is available in a machine readable form online. A part of the table is reproduced here for demonstrating the form and content of the table.

[a]The letters A–I denote stars used in photometry of P60 data, numbers 1–15 denote reference stars used in photometry of LFC images (Figure A.1, Section A.2.1).

[b]Relative photometry error. Values do not include an absolute photometry uncertainty of 0.16 mag in the g' band, 0.14 mag in the r' band and 0.064 mag in the i' band. Absolute photometry is derived using default P60 zero point calibrations.

do not correct for this contamination. The statistical uncertainty in magnitudes is $\sim$ 0.2 mag for P60 and $\sim$ 0.05 mag for LFC, and the systematic uncertainty is 0.16 mag in the g' band, 0.14 mag in the r' band and 0.06 mag in the i' band.

The resultant lightcurves are shown in Figures A.2, A.3 & A.4. Table A.1 provides the photometry.

A.2.2 Spectroscopy

We obtained optical and near-infrared spectra of 1RXS J1730+03 at various stages after outburst (Figure A.5). The first optical spectra were taken 13 days after the first photometric epoch. We used the Low Resolution Imaging Spectrograph on the 10 m Keck-I telescope (LRIS; Oke et al., 1995), with upgraded blue camera (McCarthy et al., 1998; Steidel et al., 2004), covering a wavelength range from 3,200 Å – 9,200 Å. We acquired more optical data 34 days after outburst, with the Double Beam Spectrograph on the 5 m Hale telescope at Palomar (DBSP; Oke & Gunn, 1982). We took 5 exposures spanning one complete photometric period, covering the 3,500 Å – 10,000 Å wavelength range. We took late time spectra covering just over one photometric period for the quiescent source with the upgraded LRIS[3]. At this epoch, we aligned the slit at a position angle of 45 degrees to cover both the target and the contaminator, $3''.4$ to its South West (Figure A.1). We also obtained low resolution J-band spectra with the Near InfraRed Spectrograph on the 10 m Keck-II telescope (NIRSPEC; McLean et al., 1998). 12 spectra of 5 minutes each were acquired, covering the wavelength region from 11,500 Å – 13,700 Å. For details of the observing set up, see the notes to Table A.5.

We analyzed the spectra using IRAF and MIDAS[4] and flux calibrated them using appropriate standards. Wavelength solutions were obtained using arc lamps and with offsets determined from sky emission lines. Figure A.6 shows an optical spectrum from each epoch, while the IR spectrum is shown in Figure A.7.

The second LRIS epoch had variable sky conditions. Here, we extracted spectra of the aforementioned contaminator. This object is also a M-dwarf, hence both target and contaminator spectra will be similarly affected by the atmosphere. We estimate the i' magnitude of the contaminator in each spectrum, and compare it to the the value measured from the LFC images to estimate and correct for the extinction by clouds.

A.2.3 X-ray and UV Observations

We observed the 1RXS J1730+03 with the X-ray telescope (XRT) and the UV-Optical Telescope (UVOT) onboard the *Swift* X-ray satellite (Gehrels et al., 2004) during UT 2009 May 3-6 for a total of about 12.5 ks. The level-two event data was processed using *Swift* data analysis threads for the XRT (Photon counting mode; PC) using the HEASARC FTOOLS[5] software package. The source was not detected in the X-ray band.

We follow the procedures outlined by Poole et al. (2007) for analyzing the UVOT data. The measured fluxes are given in Table A.3. The contaminator is within the recommended $5''$ extraction radius. Hence, flux measurements are upper limits.

Shevchuk et al. (2009) had observed 1RXS J1730+03 on UT 2006 February 9 with the *Swift* satellite as a part of investigations of unidentified ROSAT sources. They detected the source with a count rate of 0.02 counts s^{-1}. The best-fit power law has a photon index $\Gamma = 1.8 \pm 0.5$ and a 0.5–10 keV flux of 1.2×10^{-12} erg cm^{-2} s^{-1}. After converting to the ROSAT bandpass assuming the XRT model parameters,

---

[3]http://www2.keck.hawaii.edu/inst/lris/lris-red-upgrade-notes.html

[4]Munich Image Data Analysis System; http://www.eso.org/sci/data-processing/software/esomidas/

[5]http://heasarc.gsfc.nasa.gov/ftools/; Blackburn (1995)

Figure A.1. Finder chart for 1RXS J1730+03 ($\alpha = 17^{\mathrm{h}}30^{\mathrm{m}}06^{\mathrm{s}}.19$, $\delta = +03°38'18''.8$). This i' band image was acquired with the Large Format Camera (LFC) at the 5 m Hale telescope at Palomar. Stars numbered 1–15 in red are used for relative photometry in LFC data. Stars labeled A–I in blue are used for relative photometry in P60 data (Section A.2.1). The green circle shows the 5″ extraction region used for calculating UVOT fluxes. The circle includes a contaminator, 3″.4 to the South-West of the target.

Table A.3. Observation log for *Swift* ToO observations of 1RXS J1730+03.

Obs ID	Start Date & Time	Stop Time	Exposure (s)	Filter	Wavelength[a] (Å)	Magnitude	Flux	Flux (μJy)
00035571001	2006 Feb 9 16:56:43	18:42:00	1106	UVM2	2231	17.9±0.1	31±2[b]	51±3
				XRT			0.02[c]	0.06 (5 keV)
00031408001	2009 May 3 18:28:56	19:02:12	1964	UVM2	2231	20.44±0.21	3.04±0.59[b]	5.05±0.99
00031408002	2009 May 4 18:34:04	22:12:54	4896	UVW1	2634	20.14±0.10	3.47±0.32[b]	8.04±0.75
00031408003	2009 May 6 03:02:21	15:36:33	5621	UVW2	2030	21.48±0.17	1.38±0.22[b]	1.93±0.30
				XRT			<0.002[c]	<0.006 (5 keV)

[a]Effective wavelength for each filter for a Vega-like spectrum (Poole et al., 2007).

[b]Flux in the units of 10^{-17} erg cm^{-2} s^{-1} Å^{-1}.

[c]Counts s^{-1} in 0.5-10 keV.

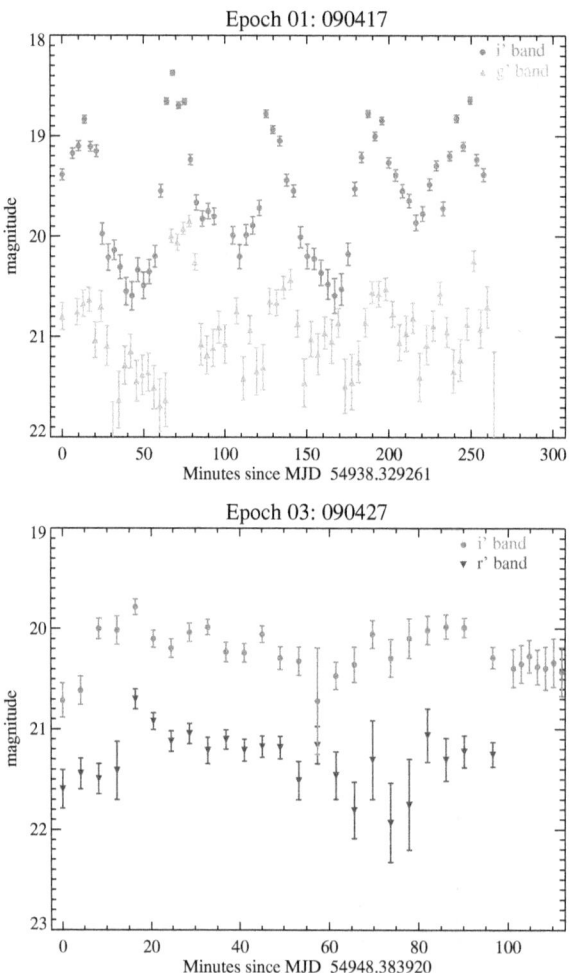

Figure A.2. P60 photometry of 1RXS J1730+03. Top panel: Epoch 1 photometry in g' and i' bands. Bottom panel: Epoch 3 photometry in r' and i' bands.

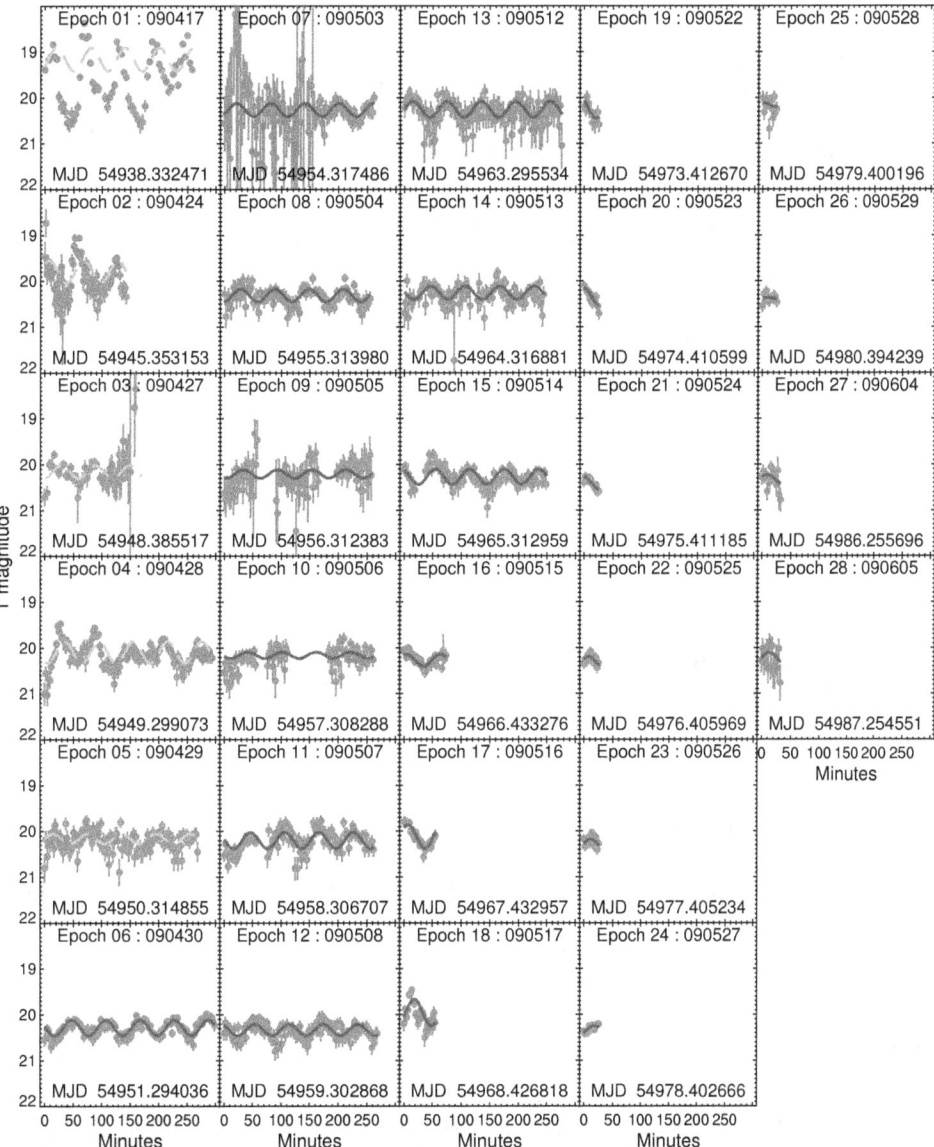

Figure A.3. i' band P60 photometry of 1RXS J1730+03, for Epochs 1 to 28. The source was in outburst in Epoch 1 and slowly faded into quiescence. The error bars denote relative photometry errors and do not include an absolute zero point error of 0.064 mag (§ A.2.1). The solid blue line denotes the best fit sinusoid to the photometric data. The dashed green line is added to show the relative phase of the source in outburst, these epochs were not used for fitting the lightcurve.

this value is approximately a factor of two lower than the archival ROSAT flux. They report a much higher UV flux in their observations, which suggests that the source was in an active state during their observations.

The column density inferred from the XRT data is low, $N_H \sim 7 \times 10^{20}$ cm^{-2}. From ROSAT data, this column density corresponds to $A_V = 0.39$, which gives $A_{UVM2} = 0.84$ (Cox, 2000). For comparison, Schlegel et al. (1998) give the Galactic dust extinction towards this direction ($l = 26.°7$, $b = 19°.7$) to be $E(B - V) = 0.141$ mag ($A_V = 0.44$), corresponding to a column density of about 8×10^{20} cm^{-2}.

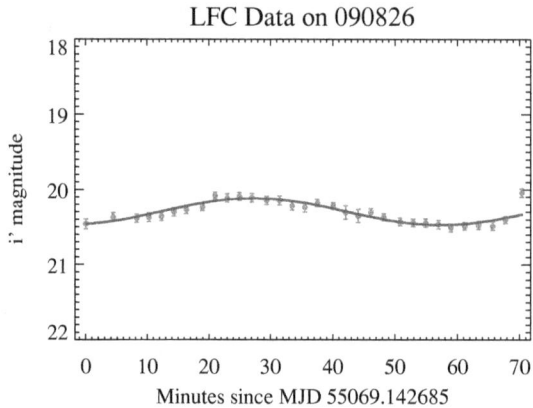

Figure A.4. i' band LFC photometry of 1RXS J1730+03 in quiescence, 131 days after we first saw high variability. The variations have the same phase and period as observed during outburst. See Figure A.3 for details.

A.3 Nature Of The Components

The optical spectra (Figure A.6) show rising flux towards the red and blue ends of the spectrum: indicative of a hot (blue) and cool (red) component. The red part of the spectrum shows clear molecular features, characteristic of late type stars. The blue component is devoid of any prominent absorption/emission features. From the overall spectral shape we infer that 1RXS J1730+03 is a CV.

A.3.1 Red Component

The red component of 1RXS J1730+03 is typical of a late type star. In Figure A.8, we compare the red side spectrum of 1RXS J1730+03 with several M-dwarfs. From the shape of the TiO bands at 7053 Å – 7861 Å, we infer that the spectral type to be M3±1. This is consistent with the relatively featureless J band spectrum (McLean et al., 2003). The spectral type indicates an effective temperature of 3400 K (Cox, 2000). The presence of a sodium doublet at 8183/8195 Å implies a luminosity class V.

We also fit the spectrum with model atmospheres calculated by Munari et al. (2005). For late type stars, these models are calculated in steps of $\Delta T = 500$ K, $\Delta \log g = 0.5$ and $\Delta[M/H] = 0.5$. We use model atmospheres with no rotational velocity ($V_{rot} = 0$ km s^{-1}) and convolve them with a kernel modeled on the seeing, slit size and pixel size. We ignore the regions contaminated by the telluric A and B bands

Table A.4. Locations of cyclotron harmonics.

Harmonic number	Measured Wavelength (Å)	Measured Frequency (Hz)	Inferred Wavelength[a] (Å)
7	3540	8.4×10^{14}	3664
6	4440	6.7×10^{14}	4275
5	5180	5.7×10^{14}	5130
4	6770	4.4×10^{14}	6413
3	8225	3.6×10^{14}	8551
2	$\cdots$	$\cdots$	12826
1	$\cdots$	$\cdots$	25653

[a]Calculated from the best fit magnetic field strength, $B = 42$ MG

(7615 Å and 6875 Å). The unknown contribution from the white dwarf was fit as a low order polynomial. We correct for extinction using $A_V = 0.39$ from X-ray data (Section A.2.3). To measure log g, we use the spectrum in the 8,000 Å – 8,700 Å region, which is expected to have fairly little contamination from the blue component. This region includes the Ca II lines at 8498, 8542 Å and the Na I doublet, which are sensitive to log g. We then fit the spectra in the 6,700 Å – 8,700 Å range to determine the temperature and metallicity. The best fit model has T = 3500 K, log $g = 5.0$ and solar metallicity, consistent with our determination of the spectral type.

Kolb et al. (2001) state that unevolved donors in CVs follow the spectral type–mass relation of the zero age main sequence, as the effects of thermal disequilibrium on the secondary spectral type are negligible. For a M3 star, this yields a mass of 0.38 $M_\odot$. As the secondaries evolve, the spectral type is no longer a good indicator of the mass and gives only an upper limit on the mass. The lower limit can simply be assumed to be the Hydrogen-burning limit of 0.08 $M_\odot$.

A.3.2 Blue Component

The blue component of 1RXS J1730+03 is consistent with a highly magnetic white dwarf. The blue spectrum is suggestive of a hot object, but does not show any prominent absorption/emission features. Hα is seen in emission, but other Balmer features are not detected. The spectrum (Figure A.6) shows cyclotron humps, suggesting the presence of a strong magnetic field. The polar nature of the object is supported by the absence of an accretion disc, and the transition from an active state to an off state in X-rays (Section A.1, Section A.2.3).

For analyzing the WD spectrum, we subtracted a scaled spectrum of the M-dwarf GL 694 from the composite spectrum of 1RXS J1730+03. The resultant spectrum (Figure A.9) clearly shows cyclotron harmonics. The hump seen in the J-band spectrum (Figure A.7) is also inferred to be a cyclotron harmonic. A detailed modeling of the magnetic field is beyond the scope of this work, but we use a simple model to estimate the magnetic field. We fit the cyclotron humps with Gaussians and measure the central wavelengths (Table A.4). We then fit these as a series of harmonics, and infer that the cyclotron frequency

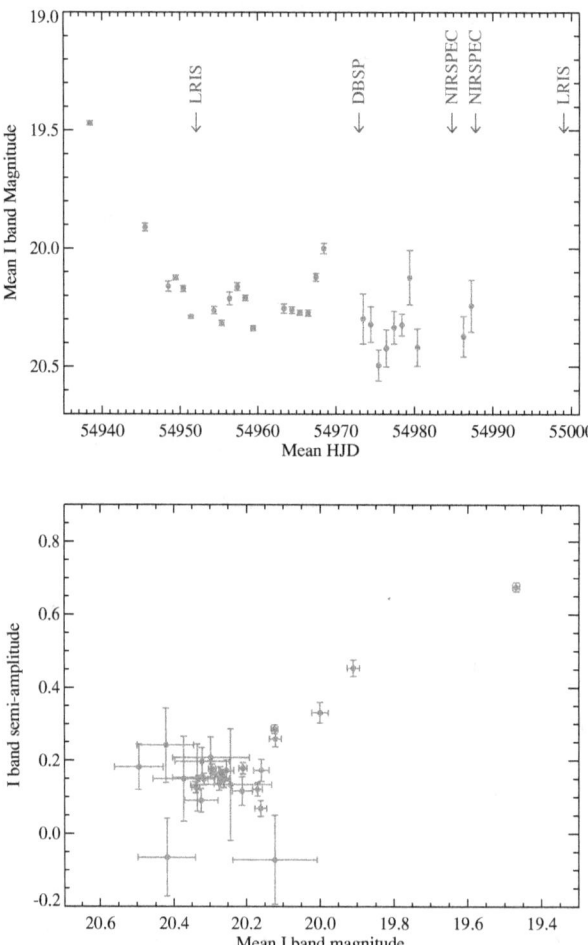

Figure A.5. Photometric evolution of 1RXS J1730+03 in i' band. Data for each epoch was fit with a sinusoid with the same period for all epochs. In contrast to Figures A.3 & A.4, the phase was allowed to vary independently for each epoch. Top: average i' magnitude for each P60 observation epoch, as a function of time. Blue arrows mark spectroscopy epochs. Bottom: semi-amplitude of sinusoidal variations as a function of the corresponding mean i' magnitudes for each epoch.

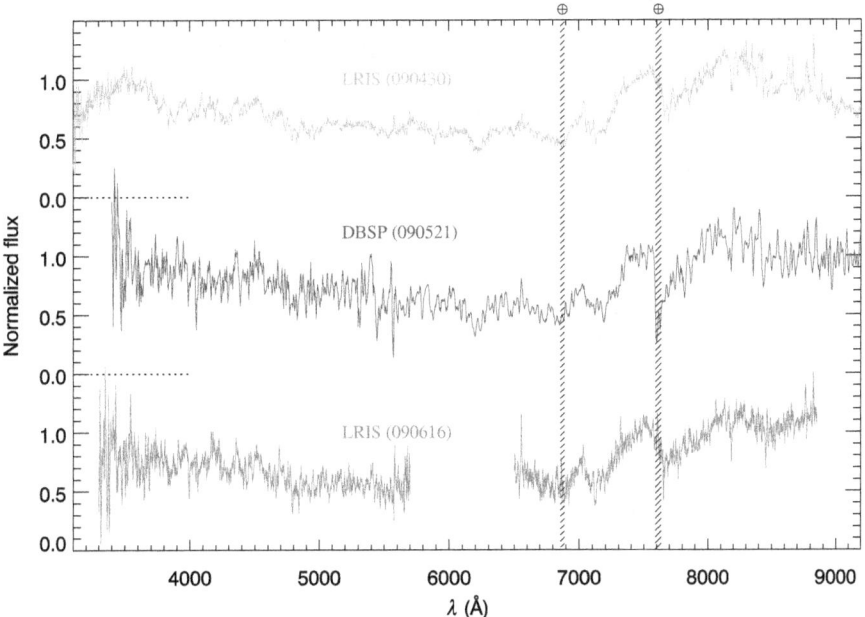

Figure A.6. Temporal evolution of 1RXS J1730+03 spectra. Some spectra are not corrected for Telluric absorption, the affected regions are diagonally hatched. The first spectrum was obtained 13 days after we measured high variability, the second after 34 days, and the lowermost spectrum after 60 days. The spectra show broad Balmer features which evolve with time. The M-dwarf features (TiO, Na I) are clearly seen at all times.

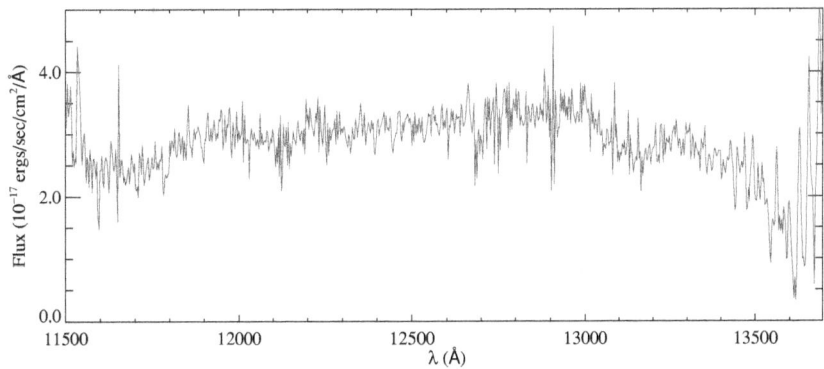

Figure A.7. Keck NIRSPEC spectrum of 1RXS J1730+03. The spectrum is flat and nearly featureless, as expected for early M stars.

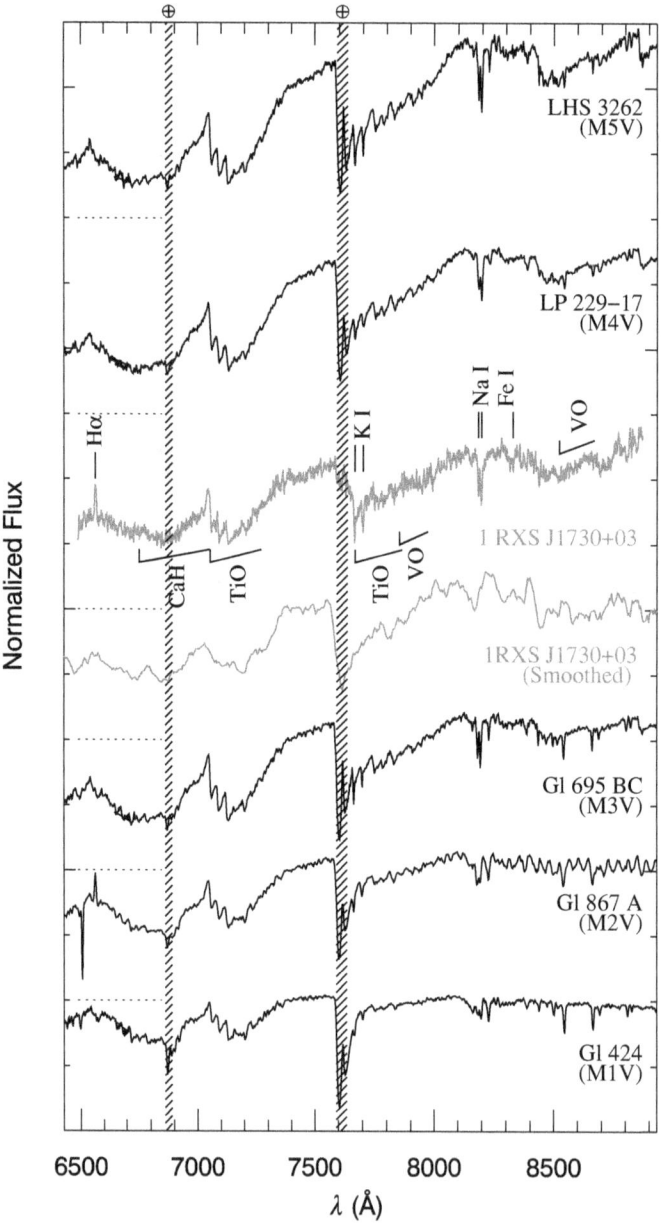

Figure A.8. Comparison of red part of 1RXS J1730+03 spectrum with M-dwarf spectra. Some spectra are not
corrected for Telluric absorption, the affected regions are diagonally hatched. Prominent bands (TiO,
CaH) and lines (Hα, Na I) are marked. Comparing the shape of the TiO bands at 7053–7861 Å and the
shape of the continuum redwards of 8200 Å, we infer that spectral type of the red component to be M3.
The presence of a sodium doublet at 8183/8195 Å implies a luminosity class V.

is $\nu_{\rm cyc} = 1.17 \times 10^{14}$ Hz. The magnetic field in the emission region is given by, $B = (\nu_{\rm cyc}/2.8 \times 10^{14}$ Hz$) \cdot 10^8$ G $= 42 \times 10^6$ G.

As a conservative error estimate, we consider the worst case scenario where our identified locations for the cyclotron humps are off by half the spacing between consecutive cyclotron harmonics. Using this, we estimate the errors on the magnetic field: $B = 42^{+6}_{-5}$ MG.

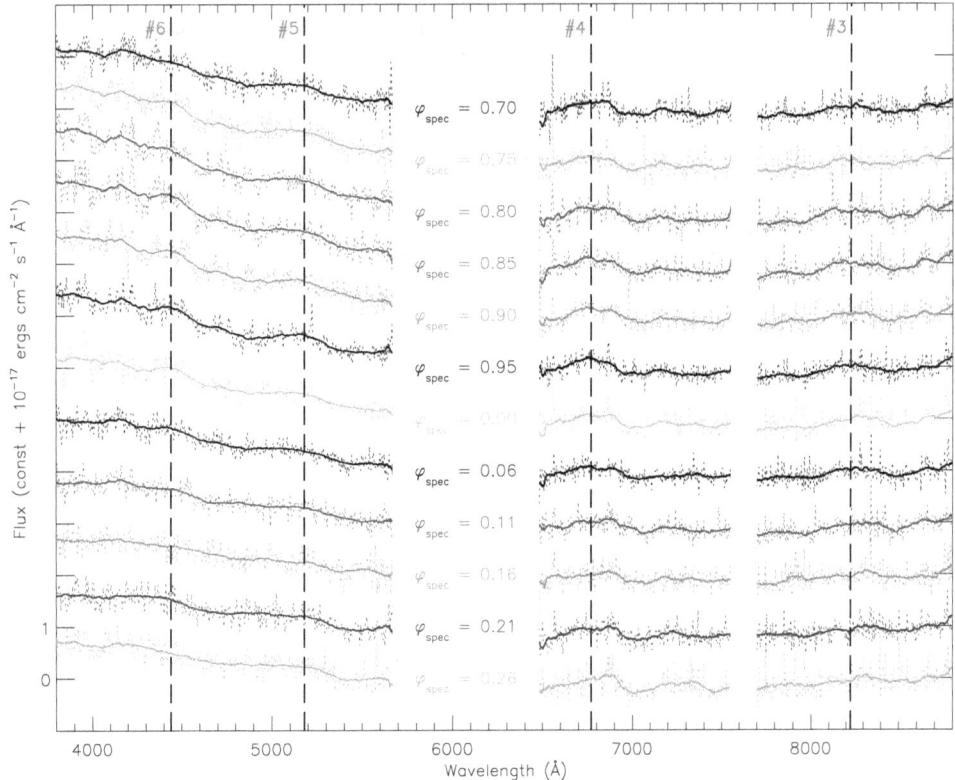

Figure A.9. Blue component spectra of 1RXS J1730+03, showing the 12 late-time LRIS exposures (Section A.2.2, Table A.5). The spectroscopic phase at mid-exposure is indicated for each spectrum. Dotted lines show spectra are obtained by subtracting a scaled spectrum of GL 694 from spectra of this binary. The overlaid solid lines are smoothed versions of the same spectra. Vertical dashed cyclotron humps at 4440 Å, 5180 Å, 6770 Å & 8225 Å. The cyclotron harmonic numbers (#3 – #6) are indicated in bold red. Two more cyclotron humps are seen in other spectra: a feature at 3540 Å (Figure A.6), and a J-band feature (Figure A.7).

A.4 System Parameters

A.4.1 Orbit

We use the best fit Munari et al. (2005) model atmosphere to measure radial velocities of the M-dwarf. We vary the radial velocity of the model, and minimize the χ^2 over the 6,500 Å – 8,700 Å spectral region, excluding the telluric O_2 bands. After a first iteration, the spectra are re-fit to account for motion of the M-dwarf during the integration time. For the 12 spectra taken at the second LRIS epoch, we also measure the radial velocity for the contaminator star on the slit, and find it to be constant. This serves as a useful test for our radial velocity measurement procedure. The barycentric corrected velocities are given in Table A.5.

We fit a circular orbit ($v_2 = \gamma_2 + K_2 \sin([2\pi(t - t_0)]/P)$) to the measured velocities. We define the superior conjunction[6] of the WD as phase 0.

The 2009 June 16 spectroscopic data (Table A.5) give an orbital period $P_{est} = 123 \pm 3$ min. We then use the photometric variability (Section A.4.2) to determine an accurate period in this range. Next, we refine the solution with velocity measurements from the other two spectroscopic epochs. The best-fit solution gives a period $P = 120.2090 \pm 0.0013$ min and $K_2 = 390 \pm 4$ km s^{-1} (Table A.6, Figure A.10).

Figure A.11 shows sections of the last epoch spectra around Hα and the Na I doublet at 8184/8195 Å. The Na I doublet clearly matches the velocity of the M-dwarf in the orbit, but the Hα emission seems to have a smaller velocity amplitude. A possible explanation for this is that the Hα emission comes from the M-dwarf surface that is closest to the WD, which may be heated by emission from the white dwarf or the accretion region.

A.4.2 Photometric Variability

The lightcurve of 1RXS J1730+03 shows clear periodicity (Figures A.2–A.4), with two peaks per spectroscopic period. Most of the photometric data was acquired in the i' band, which contains contribution from both the M-dwarf and a cyclotron harmonic from the emission region near the WD.

A Fourier transform of the data (Figure A.12) shows a strong peak at sixty minutes. We interpret this as a harmonic of the orbital period. To determine the exact period, we analyze the data as follows. As the object has a short orbital period, we convert all times to Heliocentric Julian Date for analysis. We fit a sinusoid ($m_0 + m_A \sin([2\pi(t - t_0)]/P)$) to each epoch, allowing m_0 and m_A to vary independently for each epoch, but we use the same reference time t_0 and period P for the entire fit. The mean magnitude is correlated with the amplitude (Figure A.5). Note that the amplitude measured for the sinusoidal approximation for each epoch is always less than the actual peak-to-peak variations of the source during that epoch, as expected. The source is in the active state in the first few epochs, and we exclude epochs 1 – 5 from the fit, to avoid contamination from the accretion stream and/or the accretion shock.

The best-fit solution is overplotted in blue in Figures A.3 & A.4. Since epochs 1 – 5 were not included in the fit, a sinusoid is overplotted in dashed green to indicate the expected phase of the variations. The best-fit period is 60.1059 ± 0.0005 minutes. This formally differs from the spectroscopically determined orbital period by 2.1σ. However, this error estimate includes only statistical errors. There is some, difficult to determine, systematic error component in addition, so we do not claim any significant inconsistency.

Periodic photometric variability for 1RXS J1730+03 can be explained as a combination of two effects: cyclotron emission from the accretion region and ellipsoidal modulation. The active state is characterized

---

[6]When the WD is furthest from the observer along the line of sight.

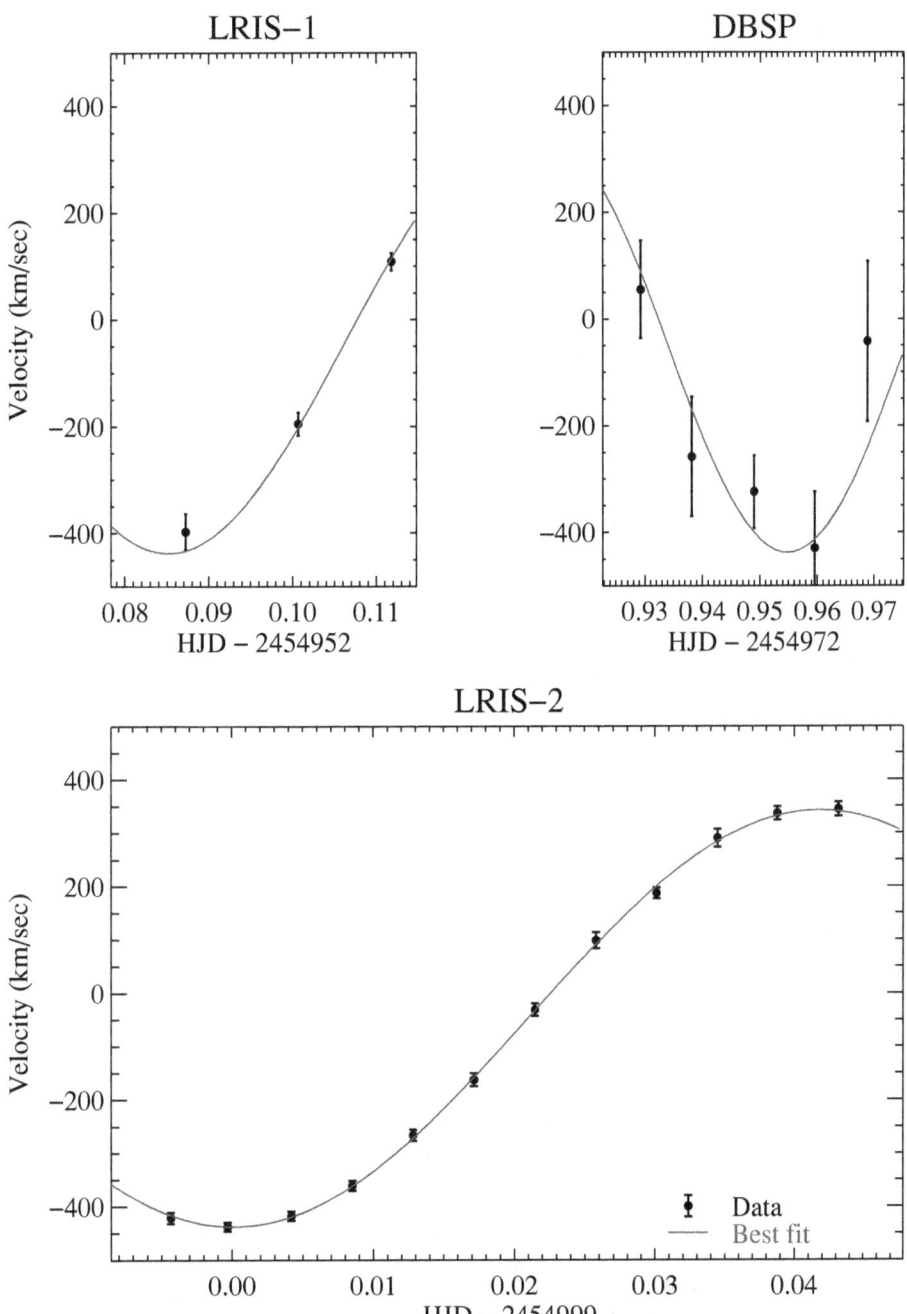

Figure A.10. Velocity measurements 1RXS J1730+03. The best-fit solution gives $P = 120.2090 \pm 0.0013$ min, $\gamma_2 = -48 \pm 5$ km s^{-1} and $K_2 = 390 \pm 4$ km s^{-1} (Section A.4.1)

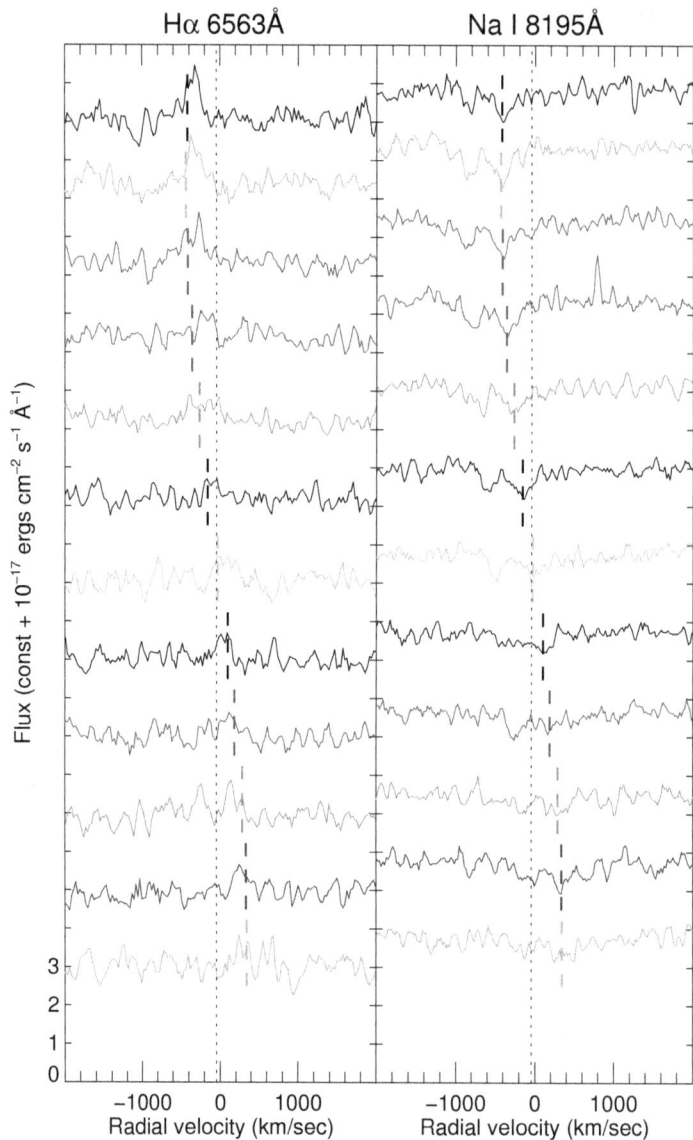

Figure A.11. Velocity modulation of Hα and the 8184/8195 Å Na I doublet. The spectra are offset by 2 × 10^{-17} erg cm^{-2} s^{-1} for clarity. The wavelengths are converted into velocities using the rest wavelengths 6563 Å and 8195 Å respectively. The vertical dotted line marks the radial velocity of the binary barycenter. The short dashed lines mark the radial velocities measured by fitting the complete spectrum. The Na I absorption follows the radial velocity of the M-dwarf, but Hα seems to have a smaller velocity amplitude.

Table A.5. Radial velocity of the M-dwarf.

Heliocentric JD	Exposure time (s)	Barycentri Radial Velocity (km s^{-1})	Instrument
2454952.08721	1020	-398 ± 34	LRIS[a]
2454952.10067	1020	-195 ± 22	LRIS[a]
2454952.11179	600	109 ± 16	LRIS[a]
2454972.92923	600	55 ± 92	DBSP[b]
2454972.93818	900	-258 ± 112	DBSP[b]
2454972.94896	900	-324 ± 69	DBSP[b]
2454972.95956	900	-430 ± 105	DBSP[b]
2454972.96875	600	-42 ± 150	DBSP[b]
2454998.99566	210	-421 ± 11	LRIS[c]
2454998.99969	300	-438 ± 9	LRIS[c]
2454999.00418	300	-417 ± 9	LRIS[c]
2454999.00850	300	-361 ± 9	LRIS[c]
2454999.01283	300	-266 ± 10	LRIS[c]
2454999.01715	300	-163 ± 12	LRIS[c]
2454999.02147	300	-32 ± 12	LRIS[c]
2454999.02580	300	98 ± 15	LRIS[c]
2454999.03012	300	186 ± 10	LRIS[c]
2454999.03448	300	290 ± 17	LRIS[c]
2454999.03880	300	336 ± 13	LRIS[c]
2454999.04317	300	344 ± 13	LRIS[c]

[a] Low Resolution Imaging Spectrograph on the 10 m Keck-I telescope (Oke et al., 1995), with upgraded blue camera (McCarthy et al., 1998; Steidel et al., 2004). Settings: Blue side: 3,200 Å – 5,760 Å, grism with 400 grooves/mm, blaze 3,400 Å, dispersion 1.09 Å pixel^{-1}, $R \sim 700$. Red side: 5450 Å–9250 Å, grating with 400 grooves/mm, blaze 8,500 Å, dispersion 1.16 Å pixel^{-1}, $R \sim 1,600$. Dichroic: 5,600 Å, slit: 1″.0.

[b] Double Spectrograph on the 5 m Hale telescope at Palomar (Oke & Gunn, 1982). Settings: Blue side: 3,270 Å–5700 Å, grating with 600 lines/mm, blaze 4,000 Å, dispersion 1.08 Å pixel^{-1}, $R \sim 1,300$. Red side: 5,300 Å–10,200 Å, grating with 158 grooves/mm, blaze 7,500 Å, dispersion 4.9Å pixel^{-1}, $R \sim 600$. Dichroic: D55 (5500 Å), slit: 1″.5.

[c] Low Resolution Imaging Spectrograph on the 10 m Keck-I telescope, with upgraded blue and red cameras; http://www2.keck.hawaii.edu/inst/lris/lris-red-upgrade-notes.html. Settings: Blue side: 3,300 Å – 5,700 Å, grism with 400 grooves/mm, blaze 3,400 Å, dispersion 1.09 Å pixel^{-1}, $R \sim 700$. Red side: 6,760 Å – 8,800 Å, grating with 831 grooves/mm, blaze 8200 Å, dispersion 0.58 Å pixel^{-1}, $R \sim 1,800$. Dichroic: 5,600 Å, slit: 1″.0.

Table A.6.   Orbital velocity parameters of the M-dwarf.

Parameter	Value
γ_2 (km s^{-1})	-48 ± 5
K_2 (km s^{-1})	390 ± 4
t_0 (HJD)	54998.9375 ± 0.0003
P (min)	120.2090 ± 0.0013

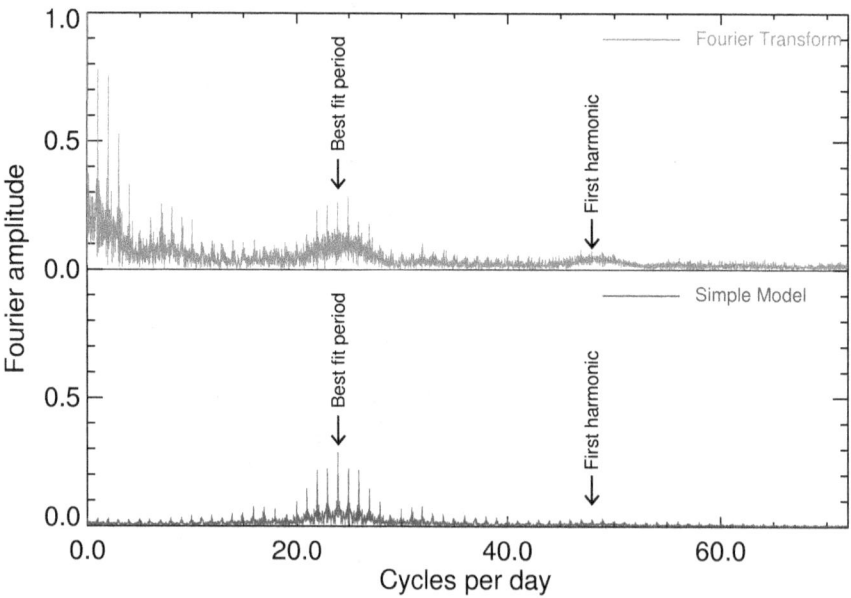

Figure A.12. Upper panel: Fourier transform of all i' data. Lower panel: expected Fourier transform for a pure sinusoidal variations with a period of 60.11 min, obtained by scaling and convolving the Fourier transform
of the window function with a delta function corresponding to the best-fit period.

by a higher mass transfer rate from the donor to the WD, and results in higher cyclotron emission. This emission is beamed nearly perpendicular to the magnetic field lines, creating an emission fan beam. For high inclination systems, the observer crosses this fan beam twice, causing two high amplitude peaks per orbit (Figure A.2). In the active state when the emission is dominated by cyclotron radiation, the minimum leads the superior conjunction by $\sim 47°$.

In ellipsoidal modulation, and the photometric minima coincide with the superior conjunction. It is observed that in quiescence, the photometric minimum leads the superior conjunction of the white dwarf by $\sim 14°$. This suggests that the 0.29 ± 0.13 mag variation is caused by a combination of ellipsoidal modulation and cyclotron emission from the accretion region.

A.4.3 Mass Ratio

The semi-amplitude of the M-dwarf radial velocity, K2, gives a lower limit on the mass of the white dwarf. For a circular orbit, one can derive from Kepler's laws that:

$$M_{1,min} = \frac{PK_2^3}{2\pi G} = 0.52 \ M_\odot \qquad (A.1)$$

Tighter constraints can be placed on the individual component masses M_1, M_2 by considering the geometry of the system. Eggleton (1983) expresses the volume radius R_2 of the secondary in terms of the mass ratio $q = M_2/M_1$:

$$\frac{R_2}{a} = \frac{0.49q^{2/3}}{0.6q^{2/3} + \ln(1 + q^{1/3})} \qquad (A.2)$$

where the separation of the components is given by $a = [P^2 G(M_1 + M_2)/4\pi^2]^{1/3}$. Thus for a fixed period P, R_2 depends only on M_2. The radius of the white dwarf is much smaller than that of the M-dwarf. Hence, the M-dwarf will eclipse the accretion region if the inclination of the system is $i < \sin^{-1}(R_2/a)$. Figures A.3, A.4 show that we do not detect any eclipses in the system.

Figure A.13 shows the allowed region for 1RXS J1730+03 in a WD mass–M-dwarf mass phase space. The orange dotted region is excluded as the orbital velocity would be greater than the measured projected velocity. The red hatched region is excluded by non-detection of eclipses. The allowed mass of the primary ranges from the minimum mass ($M_1 > 0.52 \ M_\odot$) to the Chandrashekhar limit. The mass of the secondary is bounded above by the ZAMS mass for a M3 dwarf ($M_2 = 0.38 \ M_\odot$).

For a given M_2 and a known orbital period, we can determine the radius of the secondary using Equation (A.2), and can calculate the surface gravity (log g). For $0.38 \ M_\odot \gtrsim M_2 \gtrsim 0.05 \ M_\odot$, log g ranges from 5.1 to 4.8. This is consistent with log $g = 5.0$ for the best fit M-dwarf spectrum (Section A.3.1).

The donor stars in CVs are expected to co-rotate. For 1RXS J1730+03, the highest possible rotational velocity $v \sin i$ is ~ 160 km s^{-1} for a 0.38 $M_\odot$, 0.27 $R_\odot$ M-dwarf and a 1.1 $M_\odot$ WD (Figure A.13). $v \sin i$ will be lower if the WD is heavier or if the M-dwarf is lighter. To measure rotational broadening in the spectra, we use a higher resolution ($R = 20,000$) template of the best-fit model from Zwitter et al. (2004). We take a template with zero rotation velocity and broaden it to different rotational velocities using the prescription by Gray (2005). Then we use our fitting procedure (Section A.3.1) to find the best-fit value for $v \sin i$. For this measurement, we use only the 12 relatively high resolution spectra from the second LRIS epoch (UT 2009 June 16). The weighted $v \sin i$ from the twelve spectra is 97 ± 22 km s^{-1}, but the measurements show high scatter, with a standard deviation of 54 km s^{-1}. We compared our broadened spectra with rotationally broadened spectra computed by Zwitter et al. (2004), and found that our methods systematically underestimate $v \sin i$ by ~ 20 km s^{-1}. We do not understand the reason for this discrepancy, hence do not feel confident enough to use this value in our analysis. A reliable measurement of $v \sin i$ will help better constrain the masses of the two components.

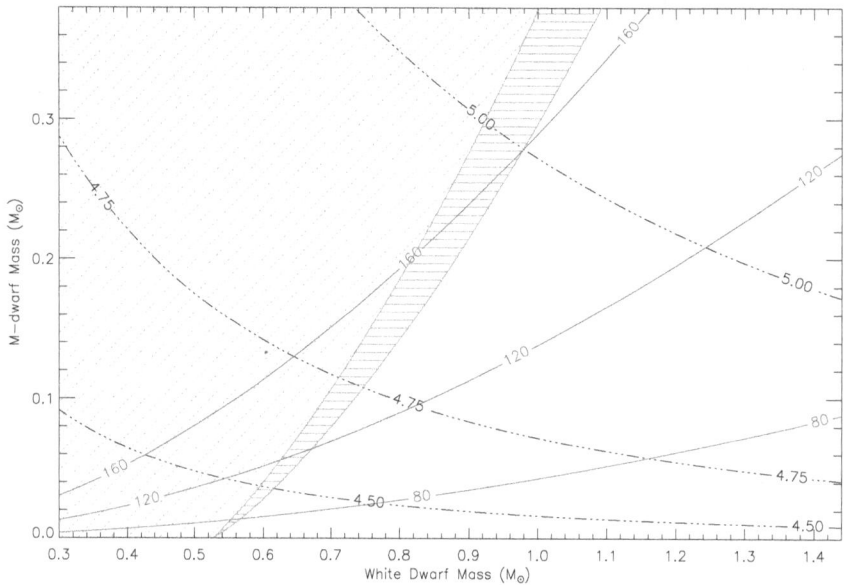

Figure A.13. The allowed range of masses (clear white region) for components of 1RXS J1730+03. The orange dotted region is excluded by the minimum inferred masses from radial velocity measurements. The red hashed region is excluded by non-detection of eclipses (Figures A.3, A.4). The black dash-dot lines show contours of constant log g. The best fit spectra yeild log $g = 5.0$. The solid blue lines are calculated contours for the rotation velocity (km s^{-1}) of the M-dwarf. A measurement of $v \sin i$ will help to constrain masses of the components.

A.4.4   Distance

We estimate the distance to 1RXS J1730+03 as follows. Our fitting procedure (Section A.3.1) corrects for extinction and separates the WD and M-dwarf components of the spectra. We correct for varying sky conditions by using a reference star on the slit. The ratio of measured flux to the flux of the best fit model atmosphere ($T = 3500$ K, $\log g = 5.0$) is,

$$\frac{f_{\text{measured}}}{f_{\text{model}}} = \left(\frac{R_2}{d}\right)^2 = (6.1 \pm 1.6) \times 10^{-23} \tag{A.3}$$

where d is the distance to the source, and R_2 is given by Equation (A.2).

For 1RXS J1730+03, the maximum mass of the M-dwarf is 0.38 $M_\odot$, and the corresponding radius is $R_2 = 0.27 \, R_\odot$. Equation (A.3) then gives $d = 800 \pm 110$ pc. This calculation assumes the largest possible M-dwarf radius, hence is an upper limit to distance. If the M-dwarf is lighter, say $0.1\ M_\odot$, we get $R_2 = 0.17\ R_\odot$, yielding $d = 500 \pm 70$ pc.

A.5   Conclusion

1RXS J173006.4+033813 is a polar cataclysmic variable, similar to known well-studied systems like BL Hyi, ST LMi and WW Hor in terms of the orbital period, magnetic field and variability between active and quiescent states. This source is notable for the highly symmetric nature and high amplitude of the double–peaked variation in the active state. This suggests a relatively high angle between the rotation and magnetic axes. Polarimetric observations of the source would help to better constrain the magnetic field geometry of the system.

Most polars are discovered due to their highly variable X-ray flux. However, we mounted a followup campaign for 1RXS J1730+03 due to its unusual optical variability properties. This suggests that current and future optical synoptic surveys, such as PTF[7] (Law et al., 2009) and LSST can uncover a large sample of polars by cross-correlating opticaly variable objects with the ROSAT catalog.

Acknowledgements

We sincerely thank the anonymous referee for detailed comments on the paper. We thank N. Gehrels for approving the Target of Opportunity observation with *Swift*, and the *Swift* team for executing the observation. We also thank V. Anupama, L. Bildsten, T. Marsh, G. Nelemans, E. Ofek and P. Szkody and for useful discussions while writing the paper.

This research has benefitted from the M, L, and T dwarf compendium housed at `DwarfArchives.org` and maintained by Chris Gelino, Davy Kirkpatrick, and Adam Burgasser.

Some of the data presented herein were obtained at the W.M. Keck Observatory, which is operated as a scientific partnership among the California Institute of Technology, the University of California and the National Aeronautics and Space Administration. The Observatory was made possible by the generous financial support of the W.M. Keck Foundation.

Facilities: PO:1.5m. Hale (LFC, DBSP), Keck:I (LRIS), Keck:II (NIRSPEC), Swift

---

[7]http://www.astro.caltech.edu/ptf

Appendix B

Transformation of a Star into a Planet in a Millisecond Pulsar Binary

M. Bailes[a,b,c,1], S. D. Bates[d] , V. Bhalerao[e], N. D. R. Bhat[a,c],
M. Burgay[f], S. Burke-Spolaor[g], N. D'Amico[f,i], S. Johnston[g],
M. J. Keith[g], M. Kramer[h,d], S. R. Kulkarni[e], L. Levin[a,g], A. G. Lyne[d],
S. Milia[i,f], A. Possenti[f], L. Spitler[a], B. Stappers[d], W. van Straten[a,c]

[a]Centre for Astrophysics and Supercomputing, Swinburne University of Technology, PO Box 218 Hawthorn, VIC 3122, Australia.

[b]Department of Astronomy, University of California, Berkeley, CA, 94720, USA.

[c]ARC Centre for All-Sky Astronomy (CAASTRO).

[d]Jodrell Bank Centre for Astrophysics, School of Physics and Astronomy, The University of Manchester, Manchester M13 9PL, UK.

[e]Caltech Optical Observatories, California Institute of Technology, MS 249-17, Pasadena, CA 91125, USA.

[f]INAF - Osservatorio Astronomico di Cagliari, Poggio dei Pini, 09012 Capoterra, Italy.

[g]Australia Telescope National Facility, CSIRO Astronomy and Space Science, P.O. Box 76, Epping NSW 1710, Australia.

[h]MPI fuer Radioastronomie, Auf dem Huegel 69, 53121 Bonn, Germany.

[i]Dipartimento di Fisica, Università degli Studi di Cagliari, Cittadella Universitaria, 09042 Monserrato (CA), Italy.

Millisecond pulsars are thought to be neutron stars that have been spun-up by accretion of matter from a binary companion. Although most are in binary systems, some 30% are solitary, and their origin is therefore mysterious. PSR J1719−1438, a 5.7 ms pulsar, was detected in a recent survey with the Parkes 64 m radio telescope. We show that it is in a binary system with an orbital period of 2.2 h. Its companion's mass is near that of Jupiter, but its minimum density of 23 g cm^{-3} suggests that it may be an ultra-low mass carbon white dwarf. This system may thus have once been an Ultra Compact Low-Mass X-ray Binary, where the companion narrowly avoided complete destruction.

A version of this chapter was first published in *Science* (Bailes et al., 2011). It is reproduced here as per the licensing policy of AAAS, http://www.sciencemag.org/feature/contribinfo/prep/lic_info.pdf, Retrieved May 6, 2012. My role was restricted to optical observations and data analysis of the counterpart.

[1]To whom correspondence should be addressed. Email: mbailes@swin.edu.au

Radio pulsars are commonly accepted to be neutron stars that are produced in the supernova explosions of their progenitor stars. They are thought to be born with rapid rotation speeds ($\sim$50 Hz) but within a few 100,000 yr slow to longer periods because of the braking torque induced by their high magnetic field strengths ($\sim10^{12}$ G). By the time their rotation periods have reached a few seconds the majority have ceased to radiate at radio wavelengths. The overwhelming majority ($\sim$ 99%) of slow radio pulsars are solitary objects. In contrast $\sim$ 70% of the millisecond pulsars (MSPs) are members of binary systems and possess spin frequencies of up to 716 Hz (Hessels et al., 2006). This is consistent with the standard model for their origin in which an otherwise dead pulsar is spun-up by the accretion of matter from a companion star as it expands at the end of its life (Bhattacharya & van den Heuvel, 1991). Through some process yet to be fully understood, the recycling not only spins up the neutron star but leads to a large reduction of the star's magnetic field strength to $B \sim 10^8$ G and usually leaves behind a white dwarf companion of typically 0.2-0.5 M$_\odot$. The lack of a compelling model for this reduction of the magnetic field strength with continuing mass accretion, and issues between the birthrates of MSPs and their putative progenitors, the low-mass X-ray binaries (LMXBs) led to an early suggestion (Grindlay & Bailyn, 1988) that accretion induced collapse of a white dwarf might form MSPs "directly" in the cores of globular clusters, and possibly in the Galactic disk.

In the standard model, the reason why some MSPs possess white dwarf companions and others are solitary is unclear. Originally it was proposed that solitary MSPs might be formed from a different channel, in which a massive ($M > 0.7$ M$_\odot$) white dwarf coalesces with a neutron star (van den Heuvel & Bonsdema, 1984). The binary pulsar-white dwarf system PSR J1141$-$6545 (Kaspi et al., 2000) is destined to merge in < 2 Gyr and thus is a potential progenitor for this scenario. At lower white dwarf masses, the final product is less clear, as the mass transfer can stabilise (Bonsema & van den Heuvel, 1985). From an observational point of view, the "black widow" MSPs may give some insights. In these systems an MSP is usually accompanied by a low-mass $\sim$ 0.02-0.05 M$_\odot$ companion in close orbits of a few hours. It was initially believed these systems might evaporate what was left of the donor star (Fruchter et al., 1988), but other examples (Stappers et al., 1998) meant that the timescales were too long.

The MSP population was further complicated by the detection of an extra-solar planetary system in orbit around the fifth MSP found in the Galactic disk, PSR B1257+12 (Wolszczan & Frail, 1992). This system has two $\sim$3 Earth-mass planets in 67- and 98-day orbits, and a smaller body of lunar mass in a 25 d orbit. The planets were probably formed from a disk of material. The origin of this disk is however the subject of much speculation, ranging from some catastrophic event in the binary that may have recycled the pulsar (Wijers et al., 1992) to ablation (Rasio et al., 1992) and supernova fall-back (Hansen et al., 2009). A large number of potential models for the creation of this system have been proposed, and are summarised in the review by Podsiadlowski (1993). Although more than another 60 MSPs ($P < 20$ ms) have been detected in the Galactic disk since PSR B1257+12, until now none have possessed planetary-mass companions.

PSR J1719$-$1438 was discovered in the High Time Resolution Universe survey for pulsars and fast transients (Keith et al., 2010). This $P = 5.7$ ms pulsar was also detected in archival data from the Swinburne intermediate latitude pulsar survey (Edwards et al., 2001). Its mean 20 cm flux density is just 0.2 mJy but at the time of discovery was closer to 0.7 mJy due to the effects of interstellar scintillation. We soon commenced regular timing of the pulsar with the Lovell 76-m telescope that soon revealed that the pulsar was a member of a binary with an orbital period of 2.17 h and a projected semi-major axis of just $a_\mathrm{p} \sin i$=1.82 ms (Figure B.1). Since then we have performed regular timing of the pulsar at the Parkes and Lovell telescopes that have enabled a phase-coherent timing solution over a one year period. There is no evidence for any statistically significant orbital eccentricity with a formal 2-σ limit of $e < 0.06$.

With these observations, we can explore the allowed range of companion masses from the binary mass function that relates the companion mass m_c, orbital inclination angle i and pulsar mass m_p to the observed projected pulsar semi-major axis a_p, orbital period P_b and gravitational constant G:

$$f(m_c) = \frac{4\pi^2}{G} \frac{(a_p \sin i)^3}{P_b^2} = \frac{(m_c \sin i)^3}{(m_c + m_p)^2} = 7.85(1) \times 10^{-10} M_\odot \tag{B.1}$$

Only a few MSPs in the Galactic disk have accurate masses (Jacoby et al., 2005; Verbiest et al., 2008; Demorest et al., 2010), and these range from 1.4–2.0 $M_\odot$. Assuming an edge-on orbit ($\sin i = 1$) and pulsar mass $m_p = 1.4$ $M_\odot$ $m_c > 1.15 \times 10^{-3}$ $M_\odot$, ie approximately the mass of Jupiter.

We can accurately determine the component separation ($a = a_p + a_c$) for the PSR J1719–1438 binary given the observed range of MSP masses using Kepler's third law and because, $m_c \ll m_p$, to a high degree of accuracy,

$$a = 0.95 R_\odot \left(\frac{m_p}{1.4\ M_\odot}\right)^{1/3} \tag{B.2}$$

making it one of the most compact radio pulsar binaries. For large mass ratios, the Roche Lobe radius of the companion (Paczynski, 1971) is well approximated by

$$R_L = 0.462a \left(\frac{m_c}{m_c + m_p}\right)^{1/3} \tag{B.3}$$

and dictates the maximum dimension of the companion star. For $i = 90^o$ and a 1.4 $M_\odot$ neutron star, the minimum $R_L = 2.8 \times 10^4$ km, just 40% of that of Jupiter. On the other hand, for a pulsar mass $m_p = 2$ $M_\odot$, and $i = 18^o$ (the chance probability of $i > 18^o$ is ~95%), then $R_L = 4.2 \times 10^4$ km.

A lower limit on the density ρ (the so-called mean density-orbital period relation, Frank et al., 1985) can be derived by combining the above equations.

$$\rho = \frac{3\pi}{0.462^3 G P_b^2} = 23\ \text{g cm}^{-3} \tag{B.4}$$

This density is independent of the inclination angle and the pulsar mass and far in excess of that of Jupiter or the other gaseous giant planets whose densities are < 2 g cm^{-3}.

The mass, radius and hence the nature of the companion of PSR J1719−1438 are critically dependent upon the unknown angle of orbital inclination. After PSR J1719−1438, PSR J2241−5236 (Keith et al., 2011) has the smallest mass function of the other binary pulsars in the Galactic disk, albeit 1000 times larger (Figure B.2). The $\sin^3 i$-dependence of the mass function could mean that PSR J1719−1438 is a physically similar system, but just viewed face-on. This would require an inclination angle of just $i = 5.7^o$, for which the chance probability is 0.5%. The only binary pulsar with a similar orbital and spin period is PSR J2051−0827, but the inclination angle required for mass function equivalence in this case has a chance probability of only 0.1%. Of course, as the known population of black widow systems increases, we will eventually observe examples of face-on binaries that mimic those with planetary-mass companions. The current distribution of mass functions among the known population is such that this is still unlikely.

If the pulsar were energetic and the orbit edge-on, we might hope to detect orbital modulation of the companion's light curve in the optical because the pulsar heats the near-side. Our pulsar timing indicates the pulsar's observed frequency derivative $\dot{\nu}$ is $-2.2(2) \times 10^{-16}$ s^{-2}, not atypical of MSPs. However $\dot{\nu}$ is only an upper limit on the intrinsic frequency derivative (Camilo et al., 1994) ($\dot{\nu_i}$), which is related to the pulsar's distance d and transverse velocity V_T by the Shklovskii relation.

$$\dot{\nu} = \dot{\nu}_i - \nu V_T^2/(dc) \tag{B.5}$$

MSPs have relatively high velocities (Toscano et al., 1999) of 50-200 km s^{-1}. At the nominal distance of the pulsar from its dispersion measure (1.2 kpc; Cordes & Lazio, 2002) it would take an MSP transverse velocity of only 100 km s^{-1} for almost all of the observed $\dot{\nu}$ to be caused by the proper motion of the pulsar.

In the case of negligible proper motion, we can derive the most optimistic impact of the pulsar's radiation for optical detectability by assuming isotropic pulsar emission, a companion albedo of unity, that the companion is a blackbody, and that the orbit is edge-on, thus maximising the illuminated region of the companion. The spin-down energy of a pulsar is $\dot{E} = -4\pi^2 I \nu \dot{\nu}$, where I is the moment of inertia of a neutron star and ν is the spin frequency. We find a maximum effective temperature of 4500 K and a peak R-band magnitude of $26-28$, depending upon the assumed 1.2(3) kpc distance to the pulsar (Cordes & Lazio, 2002) and the unknown radius of the companion, which we assume is close to the Roche Lobe radius.

We observed the field surrounding PSR J1719$-$1438 with the Keck 10-m telescope in the g, R and I bands using the LRIS instrument. If the binary was a face-on analogue of PSR J2051$-$0827 we might expect to see a star at the location of the pulsar because the R-band magnitude of the binary pulsar companion in the PSR J2051$-$0827 system is $R \sim 22.5$ (Stappers et al., 1999) and it is at a similar distance d from the Sun. The spin-down luminosity of PSR J1719$-$1438 is however only 0.4 L$_\odot$ which is about 30% of that of PSR J2051$-$0827, and a face-on orbit would mean only half of the bright side of the companion was ever visible. This would mean the expected R-band magnitude would be reduced to $R \sim 24.5$, however at the position of the pulsar there is no visible companion down to a 3-sigma limiting magnitude of R=25.4 (1250s), g=24.1 (1000s) and I=22.5 (1000s) at the anticipated maximum light, where the values in parentheses indicate the integration times (Figure B.3). The magnitude limit would appear to reduce the probability that PSR J1719$-$1438 is an extremely face-on analogue of PSR J2051$-$0827, with the caveat that the assumed spin-down energy of the pulsar is still an upper limit because of Equation B.5.

We now consider the more statistically likely possibility that the orbit is nearly edge-on. In this case the relative velocity of the two constituents is > 500 km s^{-1} and could potentially lead to a solid-body eclipse for 60 s or so, or if the companion was being ablated we might see excess dispersive delays at orbital phase 0.25 when the pulsar is on the far side of the companion at superior conjunction. Ordinarily the 20 cm mean flux density of 0.2 mJy would make these effects difficult to detect, but a bright scintillation band occurred during one of our long integrations on the source, increasing the flux density sufficiently for us to assert that there are no excess delays or solid-body eclipses occurring in the system (Figure B.1). The extremely small dimension of the Roche lobe of the companion only precludes inclination angles of $i > 87^o$. Inspection of Figure B.2 shows that it is completely impossible to fit a hydrogen-rich planet such as Jupiter into the Roche Lobe of the planetary-mass companion. Although difficult, He white dwarfs might just fit if the computational models (Deloye & Bildsten, 2003) are slighty in error ($\sim 10\%$), or the orbit is moderately face on. A carbon white dwarf on the other hand can easily fit inside the Roche Lobe for any assumed inclination angle. We thus conclude that the companion star(planet) is likely to be the remains of the degenerate core of the star that recycled the pulsar, and probably comprised of He or heavier elements such as carbon.

In the standard model, this MSP would have been spun-up by the transfer of matter from a nearby companion star to near its current period. The UC LMXBs such as XTE J0929$-$314 (Galloway et al., 2002) are good potential progenitors of PSR J1719$-$1438. These systems have orbital periods of tens of minutes and higher ($\sim 10\times$) companion masses. They have also been found to exhibit Ne and O lines in their spectra (Juett et al., 2001), suggesting that their companions are not He white dwarfs. Importantly, their spin periods are comparable to that of PSR J1719−1438. As matter is transferred from the degenerate companions to the neutron star, the orbits widen and the radius of the white dwarf expands due to the inverse mass-radius relationship for degenerate objects. Deloye & Bildsten (2003) predicted how the known UC LMXBs would evolve in the future. They demonstrated that the UC LMXB companions could be comprised of either He or Carbon white dwarfs and after 5-10 Gyr might be expected to end up as binary pulsars with orbital periods of ∼1.5h.

If PSR J1719−1438 was once a UC LMXB, mass transfer would have ceased when the radius of the white dwarf became less than that of the Roche Lobe due to mass loss and out-spiral. In the models by Deloye and Bildsten, the He white dwarfs deviate from the $M^{-1/3}$ law very near to the Roche Lobe radius and approximate mass of our companion star for an edge-on orbit. On the other hand, another mass-radius relationship for Carbon white dwarfs of very low mass (Lai et al., 1991) suggests that it is Carbon white dwarfs that have $dR/dM \sim 0$ near $M = 0.0025 M_\odot$ (Figure B.2). It thus seems difficult to unambiguously determine the nature of the pulsar companion, but a scenario in which PSR J1719−1438 evolved from a Carbon white dwarf in an UC LMXB has many attractive features. It explains the compact nature of the companion, the spin period of the pulsar and the longer orbital period due to spiral out as a consequence of the mass transfer that spun-up the pulsar. PSR J1719−1438 might therefore be the descendent of an UC LMXB.

However, the question still remains: why are some MSPs solitary while others retain white dwarf companions, and some, like PSR J1719−1438 have exotic companions of planetary mass that are possibly carbon rich? We suggest that the ultimate fate of the binary is determined by the mass and orbital period of the donor star at the time of mass transfer. Giants with evolved cores that feed the neutron star at a safe ($d >$few $R_\odot$) distance leave behind white dwarfs of varying mass in circular orbits, with a tendency for the heavier white dwarfs to be accompanied by pulsars with longer pulsar spin periods ($P >$10 ms). Close systems that transfer matter before a substantial core has formed might be responsible for the black-widow MSPs. A subset of the LMXBs are driven by gravitational radiation losses, and form the ultra-compact systems during a second stage of mass transfer. Their fate is determined by their white dwarf mass and chemical composition at the beginning of this phase. High mass white dwarfs do not overflow their Roche Lobes until they are very close to the neutron star with orbital periods of a few minutes. If the orbit cannot widen fast enough to stop runaway mass transfer we will be left with a solitary MSP or possibly an MSP with a disk that subsequently forms a planetary system. Low mass white dwarf donors transfer matter at longer orbital periods and naturally cease Roche Lobe overflow near the current orbital period and implied mass of the companion of PSR J1719−1438. The rarity of MSPs with planetary-mass companions means that the production of planets is the exception rather than the rule, and requires special circumstances, like some unusual combination of white dwarf mass and chemical composition.

PSR J1719−1438 demonstrates that special circumstances can conspire during binary pulsar evolution that allows neutron star stellar companions to be transformed into exotic planets unlike those likely to be found anywhere else in the Universe. The chemical composition, pressure and dimensions of the companion make it certain to be crystallized (ie diamond).

Parameter	Value
Right Ascension (J2000) (hh:mm:ss)	17:19:10.0730(1)
Declination (J2000) (dd:mm:ss)	-14:38:00.96(2)
ν (s^{-1})	172.70704459860(3) Hz
$\dot{\nu}$ (s^{-2})	$-2.2(2) \times 10^{-16}$
Period Epoch (MJD)	55411.0
DM (pc cm^{-3})	36.766(2)
P_b (d)	0.090706293(2)
$a_\mathrm{p} \sin i$ (lt-s)	0.001819(1)
T_0 (MJD)	55235.51652439
e	< 0.06
Data Span (MJD)	55236-55586
Weighted RMS residual (μs)	15
Points in fit	343
Mean 0.73 GHz Flux Density (mJy)	0.8*
Mean 1.4 GHz Flux Density (mJy)	0.2
Derived parameters	
Characteristic Age (Gyr)	>12.5
B (G)	$<2 \times 10^8$
Dispersion Measure Distance (kpc)	1.2 (3)
Spin-down Luminosity $L_\odot$	$<0.40(4)$

* Derived from a single observation.

Note. Differencing of two summed images at the expected maximum and minimum light of the companion also failed to reveal any modulation of flux from any potential candidates near the nominal pulsar position.

Acknowledgements.

The Parkes Observatory is part of the Australia Telescope which is funded by the Commonwealth of Australia for operation as a National Facility managed by CSIRO. This project is supported by the ARC Programmes under grants DP0985270, DP1094370 & CE110001020. Access to the Lovell telescope is funded through an STFC rolling grant. Keck telescope time is made available through a special collaborative program between Swinburne University of Technology and Caltech. We are grateful to J Roy and Y Gupta for early attempts to obtain a pulsar position with the GMRT.

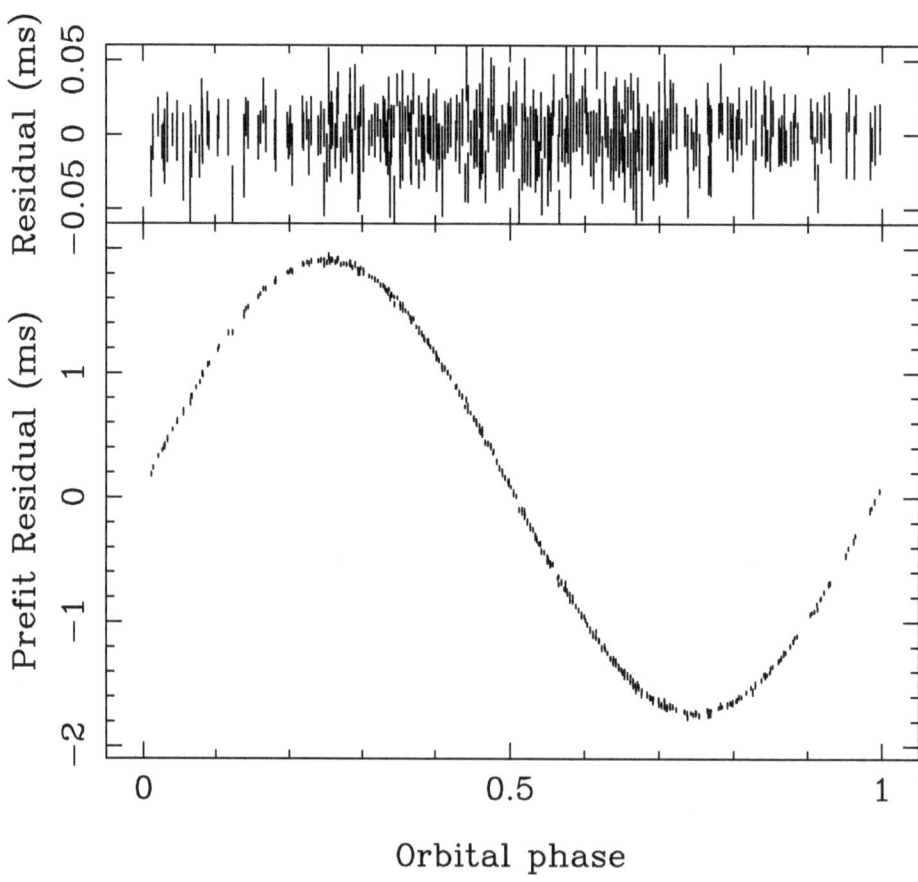

Figure B.1.   Upper panel: Pulse timing residuals for PSR J1719−1438 as a function of orbital phase using the
ephemeris in Table 1. Lower panel: Residuals after setting the semi-major axis to zero to demonstrate
the effect of the binary motion. There is no significant orbital eccentricity. At superior conjunction
(orbital phase 0.25) there is no evidence for solid-body eclipses or excess dispersive delays. The arrival
times and ephemeris are provided in the supporting online material.

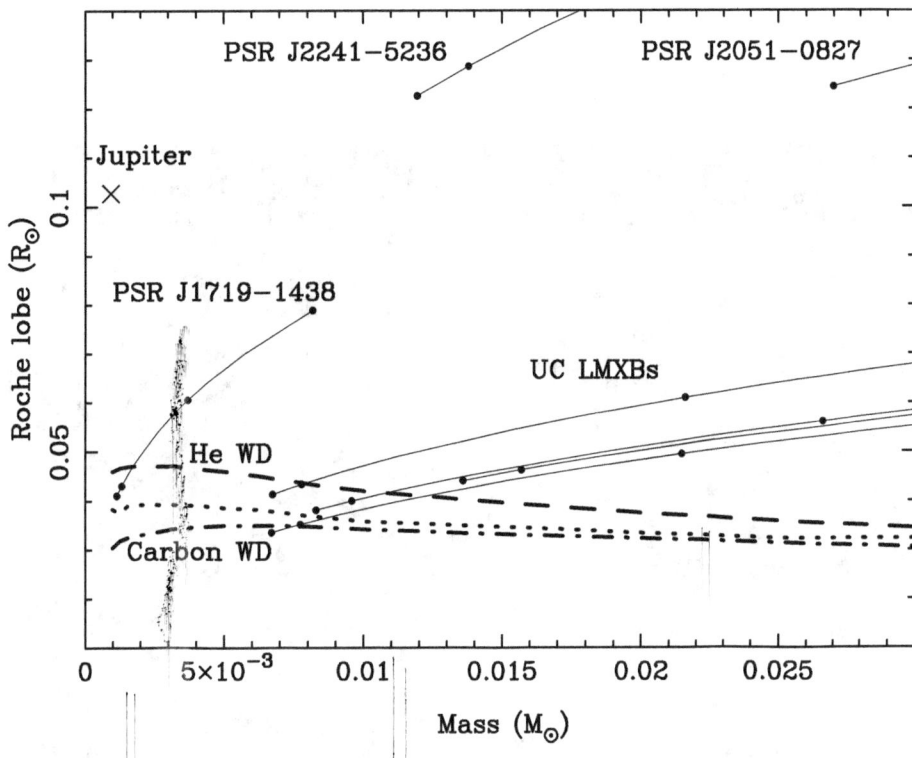

Figure B.2. The locus of the companion mass and Roche Lobe radii for PSR J1719−1438, selected ultra-compact LMXBs and black widow millisecond pulsars for different assumed orbital inclinations. The minimum companion mass and Roche Lobe radii correspond to $i = 90^o$ and a pulsar mass of 1.4 M$_\odot$. As the unknown angle of inclination decreases, the companion mass and radius increase, becoming increasingly improbable. The bullets from lowest to highest mass represent the minimum $(i = 90^o)$, median$(i = 60^o)$, 5% and 1% a priori probabilities that a randomly-oriented inclination would result in the mass and radii at least as high as that indicated. The zero-temperature mass-radius relations from Deloye & Bildsten (2003) are also shown for low-mass He and Carbon white dwarfs. The dotted line represents the mass-radius relation for low-mass Carbon white dwarfs computed by Lai et al. (1991). For reference the mass and radius of Jupiter is shown with an X.

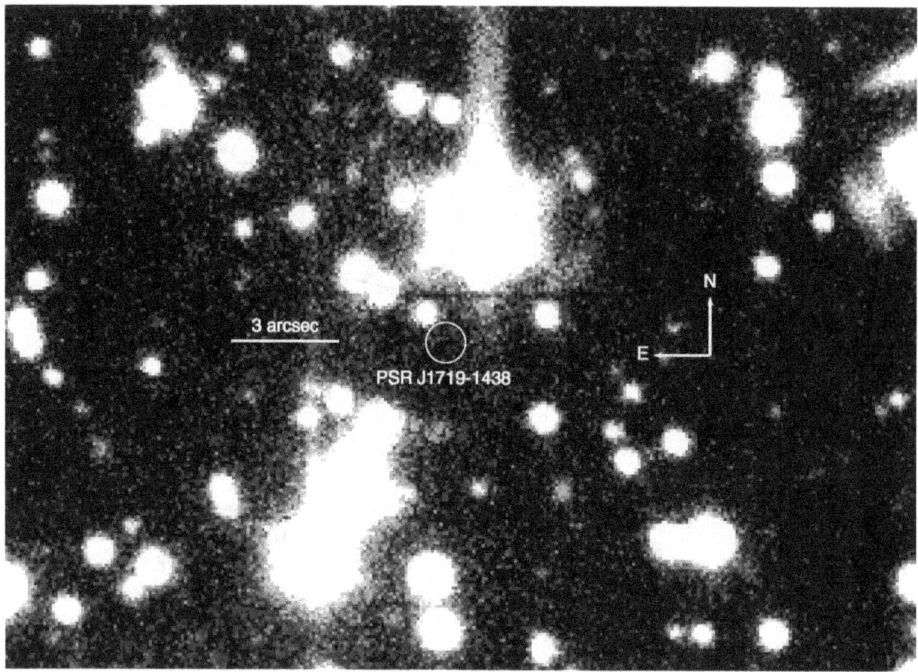

Figure B.3.   Keck/LRIS 20 minute *R*-band image centred on the location of PSR J1719−1438. The image was constructed from 5 exposures taken during the expected maximum luminosity of the companion in a total integration time of 1200s.

Bibliography

Abdo, A. A., et al. 2010, The Astrophysical Journal Supplement Series, 187, 460

Agrawal, P. 2006, Advances in Space Research, 38, 2989

Alpar, M. A., Cheng, A. F., Ruderman, M. A., & Shaham, J. 1982, Nature, 300, 728

Arzoumanian, Z., Cordes, J. M., & Wasserman, I. 1999, The Astrophysical Journal, 520, 696

Bailes, M., et al. 2011, Science, 1

Barthelmy, S. D., et al. 2005, Space Science Reviews, 120, 143

Barziv, O., Kaper, L., Van Kerkwijk, M. H., Telting, J. H., & Van Paradijs, J. 2001, Astronomy and
  Astrophysics, 377, 925

Bessell, M. S., Castelli, F., & Plez, B. 1998, Astronomy and Astrophysics, 250, 231

Bhalerao, V. B., & Kulkarni, S. R. 2011, The Astrophysical Journal, 737, L1

Bhalerao, V. B., van Kerkwijk, M. H., Harrison, F. A., Kasliwal, M. M., Kulkarni, S. R., & Rana, V. R.
  2010, The Astrophysical Journal, 721, 412

Bhattacharya, D., & van den Heuvel, E. P. J. 1991, Physics Reports, 203, 1

Bildsten, L. 1998, The Astrophysical Journal, 501, L89

—. 2003, Radio Pulsars, 302

Blackburn, J. K. 1995, Astronomical Data Analysis Software and Systems IV, 77

Bolotnikov, A., Cook, W., Harrison, F., Wong, A.-S., Schindler, S., & Eichelberger, A. 1999, Nuclear
  Instruments and Methods in Physics Research Section A: Accelerators, Spectrometers, Detectors and
  Associated Equipment, 432, 326

Bonnet-Bidaud, J. M., & Mouchet, M. 1998, Astronomy and Astrophysics, 12, 9

Bonsema, P. F. J., & van den Heuvel, E. P. J. 1985, Astronomy and Astrophysics, 146

Brott, I., et al. 2011, Astronomy & Astrophysics, 530, A115

Camero Arranz, A., Wilson, C. A., Finger, M. H., & Reglero, V. 2007, Astronomy and Astrophysics, 473,
  551

Camilo, F., Thorsett, S. E., & Kulkarni, S. R. 1994, The Astrophysical Journal, 421, L15

Casares, J., Hernández, J. I. G., Israelian, G., & Rebolo, R. 2010, Monthly Notices of the Royal Astronomical Society, 401, 2517

Cenko, S. B., et al. 2006, Publications of the Astronomical Society of the Pacific, 118, 1396

Chabrier, G., Brassard, P., Fontaine, G., & Saumon, D. 2000, The Astrophysical Journal, 543, 216

Chaty, S. 2011, Evolution of compact binaries. Proceedings of a workshop held at Hotel San Martín, 447

Chaty, S., Zurita Heras, J. A., & Bodaghee, A. 2010, eprint arXiv:1012.2318

Christensen, F. E., Hornstrup, A., Westergaard, N. J., Schnopper, H. W., Wood, J., & Parker, K. 1992,
  In: Multilayer and grazing incidence X-ray/EUV optics; Proceedings of the Meeting, 160

Christensen, F. E., et al. 2011, in Society of Photo-Optical Instrumentation Engineers (SPIE) Conference
  Series, Vol. 8147, 81470U–81470U–19

Clark, D. J., Hill, A. B., Bird, A. J., McBride, V. A., Scaringi, S., & Dean, A. J. 2009, Monthly Notices of the Royal Astronomical Society: Letters, 399, L113

Clark, J. S., Goodwin, S. P., Crowther, P. A., Kaper, L., Fairbairn, M., Langer, N., & Brocksopp, C. 2002, Astronomy and Astrophysics, 392, 909

Cordes, J. M., & Lazio, T. J. W. 2002, eprint arXiv:astro-ph/0207156

Cox, A. N. 2000, Allen's astrophysical quantities

Cropper, M. 1990, Space Science Reviews, 54, 195

de Vaucouleurs, G., de Vaucouleurs, A., Corwin, Herold G., J., Buta, R. J., Paturel, G., & Fouque, P. 1991, Volume 1-3, 1

Deloye, C. J., & Bildsten, L. 2003, The Astrophysical Journal, 598, 1217

Demorest, P. B., Pennucci, T., Ransom, S. M., Roberts, M. S. E., & Hessels, J. W. T. 2010, Nature, 467, 1081

Denisenko, D. V., Kryachko, T. V., & Satovskiy, B. L. 2009, The Astronomer's Telegram

Drave, S. P., Bird, A. J., Townsend, L. J., Hill, A. B., McBride, V. A., Sguera, V., Bazzano, A., & Clark, D. J. 2012, Astronomy & Astrophysics, 539, A21

Edwards, R., Bailes, M., van Straten, W., & Britton, M. 2001, Monthly Notices of the Royal Astronomical Society, 326, 358

Eggleton, P. P. 1983, The Astrophysical Journal, 268, 368

Figueira, P., Pepe, F., Lovis, C., & Mayor, M. 2010, Astronomy and Astrophysics, 515, A106

Frank, J., King, A. R., & Raine, D. J. 1985, Accretion power in astrophysics

Freedman, W. L., et al. 2001, The Astrophysical Journal, 553, 47

Freire, P. C. C., Ransom, S. M., Bégin, S., Stairs, I. H., Hessels, J. W. T., Frey, L. H., & Camilo, F. 2008, The Astrophysical Journal, 675, 670

Fruchter, A. S., Stinebring, D. R., & Taylor, J. H. 1988, Nature, 333, 237

Galloway, D. K., Chakrabarty, D., Morgan, E. H., & Remillard, R. A. 2002, The Astrophysical Journal, 576, L137

Gehrels, N., et al. 2004, The Astrophysical Journal, 611, 1005

Giacconi, R., Gursky, H., Paolini, F., & Rossi, B. 1962, Physical Review Letters, 9, 439

Giacconi, R., et al. 1979, The Astrophysical Journal, 230, 540

Gray, D. F. 2005, The Observation and Analysis of Stellar Photospheres

Grindlay, J. E., & Bailyn, C. D. 1988, Nature, 336, 48

Hailey, C. J., et al. 2010, in Space Telescopes and Instrumentation 2010: Ultraviolet to Gamma Ray. Edited by Arnaud, Vol. 7732, 77320T–77320T–13

Hansen, B. M. S., Shih, H.-Y., & Currie, T. 2009, The Astrophysical Journal, 691, 382

Harrison, F. A., et al. 2006, Experimental Astronomy, 20, 131

Harrison, F. A., et al. 2010, in Space Telescopes and Instrumentation 2010: Ultraviolet to Gamma Ray. Edited by Arnaud, Vol. 7732, 77320S–77320S–8

Harrison, M., McGregor, D., & Doty, F. 2008, Physical Review B, 77

Harwit, M. 2003, Physics Today, 56, 38

Hellier, C. 2001, Cataclysmic Variable Stars

—. 2002, The Physics of Cataclysmic Variables and Related Objects, 261

Hessels, J., Ransom, S., Roberts, M., Kaspi, V., Livingstone, M., Tam, C., & Crawford, F. 2005, Binary Radio Pulsars, 328

Hessels, J. W. T., Ransom, S. M., Stairs, I. H., Freire, P. C. C., Kaspi, V. M., & Camilo, F. 2006, Science (New York, N.Y.), 311, 1901

Hilditch, R. W. 2001, An Introduction to Close Binary Stars, 1st edn. (Cambridge, UK: Cambridge University Press)

Hilditch, R. W., Howarth, I. D., & Harries, T. J. 2005, Monthly Notices of the Royal Astronomical Society, 357, 304

Hulleman, F., in 't Zand, J. J. M., & Heise, J. 1998, Astronomy and Astrophysics, 337, L25

in 't Zand, J. J. M. 2005, Astronomy and Astrophysics, 441, L1

Iniewski, K. 2010, Semiconductor radiation detection systems (Boca Raton FL: CRC Press/Taylor & Francis)

Jacoby, B. A., Hotan, A., Bailes, M., Ord, S., & Kulkarni, S. R. 2005, The Astrophysical Journal, 629, L113

Jahoda, K., Swank, J. H., Giles, A. B., Stark, M. J., Strohmayer, T., Zhang, W., & Morgan, E. H. 1996, Proc. SPIE Vol. 2808, 2808, 59

Jester, S., et al. 2005, The Astronomical Journal, 130, 873

Joss, P. C., & Rappaport, S. A. 1984, Annual Review of Astronomy and Astrophysics, 22, 537

Juett, A. M., Psaltis, D., & Chakrabarty, D. 2001, The Astrophysical Journal, 560, L59

Kaspi, V. M., et al. 2000, The Astrophysical Journal, 543, 321

Kaur, R., Wijnands, R., Paul, B., Patruno, A., & Degenaar, N. 2010, Monthly Notices of the Royal Astronomical Society, 402, 2388

Keith, M. J., et al. 2010, Monthly Notices of the Royal Astronomical Society, 409, 619

—. 2011, Monthly Notices of the Royal Astronomical Society, 414, 1292

Kiziltan, B., Kottas, A., & Thorsett, S. E. 2010, eprint arXiv:1011.4291

Kiziltan, B., & Thorsett, S. E. 2010, The Astrophysical Journal, 715, 335

Knigge, C., Coe, M. J., & Podsiadlowski, P. 2011, Nature, 479, 372

Kolb, U., King, A. R., & Baraffe, I. 2001, Monthly Notices of the Royal Astronomical Society, 321, 544

Kzlolu, U., Kzlolu, N., Baykal, A., Yerli, S. K., & Özbey, M. 2007, Astronomy and Astrophysics, 470, 1023

Lai, D., Abrahams, A. M., & Shapiro, S. L. 1991, The Astrophysical Journal, 377, 612

Landsman, W. B. 1993, Astronomical Data Analysis Software and Systems II, 52

Lang, D., Hogg, D. W., Mierle, K., Blanton, M., & Roweis, S. 2010, The Astronomical Journal, 139, 1782

Lattimer, J., & Prakash, M. 2005, Physical Review Letters, 94

Lattimer, J. M., & Prakash, M. 2004, Science (New York, N.Y.), 304, 536

—. 2007, Physics Reports, 442, 109

Lattimer, J. M., Prakash, M., Day, R. R., & Year, M. 2010, eprint arXiv:1012.3208, 1

Law, N. M., et al. 2009, Publications of the Astronomical Society of the Pacific, 121, 1395

Lebrun, F., et al. 2003, Astronomy and Astrophysics, 411, L141

Lin, J., Rappaport, S., Podsiadlowski, P., Nelson, L., Paxton, B., & Todorov, P. 2011, The Astrophysical Journal, 732, 70

Liu, Q. Z., van Paradijs, J., & van den Heuvel, E. P. J. 2005, Astronomy and Astrophysics, 442, 1135

—. 2006, Astronomy and Astrophysics, 455, 1165

Longair, M. S. 1992, High Energy Astrophysics

Madsen, K. K., Harrison, F. a., Mao, P. H., Christensen, F. E., Jensen, C. P., Brejnholt, N., Koglin, J., & Pivovaroff, M. J. 2009, Proceedings of SPIE, 7437, 743716

Manchester, R. N., Hobbs, G. B., Teoh, A., & Hobbs, M. 2005, The Astronomical Journal, 129, 1993

Mao, P. H., Harrison, F. A., Windt, D. L., & Christensen, F. E. 1999, Applied Optics, 38, 4766

Markwardt, C. B. 2009, Astronomical Data Analysis Software and Systems XVIII ASP Conference Series, 411

Mason, A. B., Clark, J. S., Norton, A. J., Crowther, P. A., Tauris, T. M., Langer, N., Negueruela, I., &

Roche, P. 2011a, Monthly Notices of the Royal Astronomical Society, 000, 10

Mason, A. B., Norton, A. J., Clark, J. S., Negueruela, I., & Roche, P. 2011b, Astronomy & Astrophysics, 532, A124

Mason, A. B., Norton, A. J., Clark, J. S., Roche, P., & Negueruela, I. 2010, Société Royale des Sciences de Liège, 80, 699

Massey, P., Morrell, N. I., Neugent, K. F., Penny, L. R., DeGioia-Eastwood, K., & Gies, D. R. 2012, The Astrophysical Journal, 748, 96

McCarthy, J. K., et al. 1998, Proc. SPIE Vol. 3355, 3355, 81

McConnachie, A. W., Irwin, M. J., Ferguson, A. M. N., Ibata, R. A., Lewis, G. F., & Tanvir, N. 2005, Monthly Notices of the Royal Astronomical Society, 356, 979

McLean, I. S., McGovern, M. R., Burgasser, A. J., Kirkpatrick, J. D., Prato, L., & Kim, S. S. 2003, The Astrophysical Journal, 596, 561

McLean, I. S., et al. 1998, Proc. SPIE Vol. 3354, 3354, 566

Mighell, K. J. 1999, Astronomical Data Analysis Software and Systems VIII, 172

Munari, U., Sordo, R., Castelli, F., & Zwitter, T. 2005, Astronomy and Astrophysics, 442, 1127

MunozDarias, T., Casares, J., & MartinezPais, I. G. 2005, The Astrophysical Journal, 635, 502

Negueruela, I., & Schurch, M. P. E. 2007, Astronomy and Astrophysics, 461, 631

Niemela, A., & Sipila, H. 1994, IEEE Transactions on Nuclear Science, 41, 1054

Nomoto, K. 1984, The Astrophysical Journal, 277, 791

Oke, J. B., & Gunn, J. E. 1982, Publications of the Astronomical Society of the Pacific, 94, 586

Oke, J. B., et al. 1995, Publications of the Astronomical Society of the Pacific, 107, 375

Özel, F., Psaltis, D., Narayan, R., & McClintock, J. E. 2010, The Astrophysical Journal, 725, 1918

Özel, F., Psaltis, D., Narayan, R., & Villarreal, A. S. 2012, arXiv eprint

Paczynski, B. 1971, Annual Review of Astronomy and Astrophysics, 9, 183

Paul, B., & Naik, S. 2011, Bulletin of the Astronomical Society of India, 39, 429

Pellizza, L. J., Chaty, S., & Negueruela, I. 2006, Astronomy and Astrophysics, 455, 653

Pietsch, W., Misanovic, Z., Haberl, F., Hatzidimitriou, D., Ehle, M., & Trinchieri, G. 2004, Astronomy and Astrophysics, 426, 11

Pietsch, W., Plucinsky, P. P., Haberl, F., Shporer, A., & Mazeh, T. 2006, The Astronomer's Telegram, 905, 1

Pietsch, W., et al. 2009, The Astrophysical Journal, 694, 449

Podsiadlowski, P. 1993, in Planets around pulsars; Proceedings of the Conference, ed. J. A. Phillips, S. E. Thorsett, & S. R. Kulkarni, Vol. 36, 149–165

Poole, T. S., et al. 2007, Monthly Notices of the Royal Astronomical Society, 383, 627

Quaintrell, H., Norton, A. J., Ash, T. D. C., Roche, P., Willems, B., Bedding, T. R., Baldry, I. K., & Fender, R. P. 2003, Astronomy and Astrophysics, 401, 313

Radhakrishnan, V., & Srinivasan, G. 1982, Current Science, 51, 1096

Rahoui, F., Chaty, S., Lagage, P.-O., & Pantin, E. 2008, Astronomy and Astrophysics, 484, 801

Ramsay, G., Cropper, M., Wu, K., Mason, K. O., Córdova, F. A., & Priedhorsky, W. 2004, Monthly Notices of the Royal Astronomical Society, 350, 1373

Rappaport, S., Podsiadlowski, P., Joss, P. C., Di Stefano, R., & Han, Z. 1995, Monthly Notices of the Royal Astronomical Society, 273, 731

Rappaport, S. A., & Joss, P. C. 1983, IN: Accretion-driven stellar X-ray sources (A84-35577 16-90). Cambridge and New York, 1

Rasio, F. A., Shapiro, S. L., & Teukolsky, S. A. 1992, Astronomy and Astrophysics (ISSN 0004-6361), 256

Reig, P. 2011, Astrophysics and Space Science, 332, 1

Reig, P., Negueruela, I., Fabregat, J., Chato, R., Blay, P., & Mavromatakis, F. 2004, Astronomy and Astrophysics, 421, 673

Remillard, R. A., & McClintock, J. E. 2006, Annual Review of Astronomy and Astrophysics, 44, 49

Rephaeli, Y., Nevalainen, J., Ohashi, T., & Bykov, A. M. 2008, Space Science Reviews, 134, 71

Roberts, M. S. E., Gotthelf, E. V., Halpern, J. P., Brogan, C. L., & Ransom, S. M. 2007, Proceedings of the 363. WE-Heraeus Seminar on Neutron Stars and Pulsars 40 years after the discovery. Edited by W. Becker and H. H. Huang. MPE-Report 291. ISSN 0178-0719. Published by the Max Planck Institut für extraterrestrische Physik

Rockosi, C., et al. 2010, in Ground-based and Airborne Instrumentation for Astronomy III. Edited by McLean, Vol. 7735, 77350R–77350R–11

Schlegel, D. J., Finkbeiner, D. P., & Davis, M. 1998, The Astrophysical Journal, 500, 525

Schwab, J., Podsiadlowski, P., & Rappaport, S. 2010, The Astrophysical Journal, 719, 722

Shevchuk, A. S., Fox, D. B., Turner, M., & Rutledge, R. E. 2009, The Astronomer's Telegram

Shklovskii, I. S. 1970, Soviet Astronomy, 13

Shporer, A., Hartman, J., Mazeh, T., & Pietsch, W. 2006, The Astronomer's Telegram, 913, 1

Sidoli, L. 2011, Advances in Space Research, 48, 88

Simcoe, R. A., Metzger, M. R., Small, T. A., & Araya, G. 2000, American Astronomical Society, 32

Smith, D. M. 2004, The Astronomer's Telegram

Stappers, B. W., Bailes, M., Manchester, R. N., Sandhu, J. S., & Toscano, M. 1998, The Astrophysical Journal, 499, L183

Stappers, B. W., van Kerkwijk, M. H., Lane, B., & Kulkarni, S. R. 1999, The Astrophysical Journal, 510, L45

Steidel, C. C., Shapley, A. E., Pettini, M., Adelberger, K. L., Erb, D. K., Reddy, N. A., & Hunt, M. P. 2004, The Astrophysical Journal, 604, 534

Steiner, A. W., Lattimer, J. M., & Brown, E. F. 2010, The Astrophysical Journal, 722, 33

Sunyaev, R. A., Grebenev, S. A., Lutovinov, A. A., Rodriguez, J., Mereghetti, S., Gotz, D., & Courvoisier, T. 2003, The Astronomer's Telegram, 190, 1

Takahashi, T., et al. 2007, Publications of the Astronomical Society of Japan

Tauris, T. M., Langer, N., & Kramer, M. 2011, Monthly Notices of the Royal Astronomical Society, 416, 2130

Timmes, F. X., Woosley, S. E., & Weaver, T. A. 1996, The Astrophysical Journal, 457, 834

Torres, G. 2010, The Astronomical Journal, 140, 1158

Toscano, M., Sandhu, J. S., Bailes, M., Manchester, R. N., Britton, M. C., Kulkarni, S. R., Anderson, S. B., & Stappers, B. W. 1999, Monthly Notices of the Royal Astronomical Society, 307, 925

Tousey, R., Watanabe, K., & Purcell, J. 1951, Physical Review, 83, 792

Ubertini, P., et al. 2003, Astronomy and Astrophysics, 411, L131

Valentim, R., Rangel, E., & Horvath, J. E. 2011, Monthly Notices of the Royal Astronomical Society, 414, 1427

van den Heuvel, E. P. J., & Bonsdema, P. T. J. 1984, Astronomy and Astrophysics (ISSN 0004-6361), 139

van der Meer, A., Kaper, L., van Kerkwijk, M. H., Heemskerk, M. H. M., & van den Heuvel, E. P. J. 2007, Astronomy and Astrophysics, 473, 523

van Dokkum, P. G. 2001, Publications of the Astronomical Society of the Pacific, 113, 1420

van Kerkwijk, M. H., Breton, R. P., & Kulkarni, S. R. 2011, The Astrophysical Journal, 728, 95

van Kerkwijk, M. H., van Paradijs, J., Zuiderwijk, E. J., Hammerschlag-Hensberge, G., Kaper, L., &

Sterken, C. 1995, Astronomy and Astrophysics

Verbiest, J. P. W., et al. 2008, The Astrophysical Journal, 679, 675

Voges, W., et al. 1999, Astronomy and Astrophysics

Walborn, N. R., & Fitzpatrick, E. L. 1990, Publications of the Astronomical Society of the Pacific, 102, 379

Warner, B. 1995, Camb. Astrophys. Ser., 28

Weisskopf, M. C., Tananbaum, H. D., Van Speybroeck, L. P., & O'Dell, S. L. 2000, Proc. SPIE Vol. 4012, 4012, 2

Wijers, R. A. M. J., van den Heuvel, E. P. J., van Kerkwijk, M. H., & Bhattacharya, D. 1992, Nature, 355, 593

Wilson, C. A., Finger, M. H., Harmon, B. A., Chakrabarty, D., & Strohmayer, T. 1998, The Astrophysical Journal, 499, 820

Wilson, C. A., Weisskopf, M. C., Finger, M. H., Coe, M. J., Greiner, J., Reig, P., & Papamastorakis, G. 2005, The Astrophysical Journal, 622, 1024

Wolszczan, A., & Frail, D. A. 1992, Nature, 355, 145

Wolter, H. 1952a, Annalen der Physik, 445, 286

—. 1952b, Annalen der Physik, 445, 94

York, D. G., et al. 2000, The Astronomical Journal, 120, 1579

Zhang, C. M., et al. 2011, Astronomy & Astrophysics, 527, A83

Zwitter, T., Castelli, F., & Munari, U. 2004, Astronomy and Astrophysics, 417, 1055

www.ingramcontent.com/pod-product-compliance
Lightning Source LLC
Chambersburg PA
CBHW081124170526

45165CB00008B/2537